BEYOND REPAIR

Beyond Repair

The Ecology of Capitalism

by Barry Weisberg

Beacon Press Boston

Library of Congress catalog card number: 75-156455
International Standard Book Number: 0-8070-0536-3
Beacon Press books are published under the auspices
of the Unitarian Universalist Association
Published simultaneously in Canada
by Saunders of Toronto, Ltd.

Printed in the United States of America

To be sure

the present moment and
the one immediately ahead
are horrifying and will
become increasingly so. . . .
But the moment further on
in the distance will be
utterly brilliant. . . . .
The belly of the earth
is still full of eggs . . .

Nikos Kazantzakis

A century devoted to the rationality of technique was also a century so irrational as to open in every mind the real possibility of global destruction. It was the first century in history which presented to sane and sober minds the fair chance that the century might not reach the end of its span. It was a world half convinced of the future death of our species yet half aroused by the apocalyptic notion that an exceptional future still lay ahead of us. So it was a century which moved with the magnificent display of power into directions it could not comprehend. The itch was to accelerate — the metaphysical direction unknown. — Norman Mailer, *Of a Fire on the Moon*

DEDICATION

For those who look toward that exceptional future. Waiting, thus, in the belly of the whale, it is a rough ride. Rougher than to the moon by far. For the end of all our explorations is not the moon but the earth, not alone in a machine but together with other people. Jonah waits. Let the whale brood; the time approaches.

Contents

One: Introduction: The Power to Destroy and the Power to Create 1

Two: From Past to Present 13

Three: The Structure of Social and Ecological Responsibility 41

Four: The War Machine 77

Five: Oiling the Machine: Automobiles and Petroleum 98

Six: The Conditions of Liberation 146

Notes 185

Index 189

BEYOND REPAIR

I

Introduction: The Power to Destroy and the Power to Create

We have broken with the naïve evidence of life. We walk the streets of America asleep to the poisons which consume our body and mind. One month we are concerned about the war, the next DDT. Month after month of horror stories. To all this we have come to adapt.

Yet we do not truly experience the war or DDT. No more than we truly experience the vast natural foundation which is our heritage. Our mind has escaped its roots in the body of this world. We are estranged from our natural roots, the air, land, water, and other animals. We are separated from one another and from the world of which we are part. It is the separation[1] between man and nature, men and women, mind and body, rich and poor, white and black, mine and thine, past, present, and future. It is rooted in the social division of labor and the formation of classes.

The destruction of the biosphere as a suitable habitat for life as we know it is not simply a matter of too many people or that some people consume too much of the others' share, but is rooted in the systemic imbalance between the capitalist organization of society and the life-sustaining capacity of this planet. The daily imposition of capitalism upon the functioning of the life support system exacerbates geometrically the future of biological evolution.

The profound imbalance in the natural world is rooted in the imbalance America has generated in the social sphere. Entire generations contaminated and terrorized by Hiroshima and Nagasaki. The vast sea of malnutrition and hunger resulting from the American economic colonization of the world. The hierarchy of race and sex. Seven percent of the world's population in thc United

States busily consuming between 50 and 70 percent of the world's resources.

This crisis, at once social and biological, is not simply a matter of "new priorities," "overpopulation," or "pollution." The poisoning and imbalance generated within the natural web cannot be truly comprehended unless at the same time one examines the inhuman forms of social organization responsible for this unnatural condition.

It is foolish to focus attention upon Santa Barbara or auto emissions without at the same time examining the structure of social, political, and economic relationships responsible for these events. Muckraking individuals or particular decisions are diversions from the underlying political economy of petroleum and automobiles which is at the root of the problem. One can no more understand events in isolation in the social context than in the natural dimension. It is the organization behind the parts, biologically and socially, which is the proper subject, ecologically and politically, of our current peril. As the relationships among organisms are the foundations of any ecosystem, in the modern political state the relationships between the entities are as important as the entities themselves. Thus the political framework for an examination of our ecological crisis is derived from the basic configurations of ecological science — the relationships between organisms and their environment.

We must discover anew the structure of the perceived world. This structure is buried under the sedimentations of knowledge propagated by the schools, churches, families, and the multitude of institutions dedicated to bringing up "Americans." Much in the sense that Marcel Duchamp suggested by the comment, "The spectator makes the picture," the social critic today must discover anew the foundations of social reality.

Today that reality as a global system is determined first and foremost by the United States. The imbalance generated within the present-day biological and social organization has to date its ultimate expression in the United States.

ii

To grasp this we must begin by realizing that interdependence is a necessary condition of all life. This continual process of exchange requires a balance between input and output, give and take among all organisms. Every organism both partakes of and contributes to the energy balance of all life. Because the production and consumption of every living organism have a function, we are completely dependent upon all other forms of life. Nothing in the world is "useless," without purpose or design. The human hierarchy over nature — which results from hierarchy within human culture — has no justification in the priorities of evolution. As Loren Eiseley has pointed out, "It pays to know there is just as much future as there is past. The only thing it doesn't pay to be sure of is man's part in it."[2]

The separation of man over and above nature is a false separation. It reflects the enforced hierarchy which is the basis of the social and ecological conditions of our day.

The way out, the healing of the world's body, is found in the nascent logic of our own perception. As Merleau-Ponty suggests, "We must take literally what vision teaches us; namely that through it we come into contact with the sun and the stars, that we are everywhere all at once."[3]

In fact we are billion-year-old carbon. Stardust. We are the creatures first venturing up out of the waters, the explosion of flowers, the coming of mountain ranges, and the receding ice. We are all potential fossils.

All life is coterminous with matter, and thus with all other life. Life is a single organic fabric with an infinite number of threads that tie together the several billion years of evolutionary process. They inextricably link the hundred million years of man to the four billion years of earth, to the universe without beginning or end.

In a matter of decades the United States has begun to unravel these threads. Twentieth century America has altered the chemistry of the atmosphere and begun to empty the vast oceanic vat. Behind the glossy, earth-bound image of the moon lies the abandoned debris of Apollo. Women now fear to breastfeed their children because of the harmful concentrations of DDT in the milk. A fate shared by both polar bears and penguins.

The manipulation and exploitation of the life support system and the production of hazardous chemicals are today compounded by arrogance upon ignorance. This imbalance is the result of a massive system of irrationality which literally blinds its inhabitants to the experience of a decaying world. This imbalance is reflected daily by the concentrations of mercury, radiation, and the constant parade of other hazardous substances. This imbalance is reflected in the congested urban center, the constant noise, and the banality of food — in taste and appearance. Originally such imbalances affected only local organisms and the immediate surroundings. Today, however, when the planetary social organization lends support to perpetuate this imbalance — the military complex, the massive production of chemical effluents in major industries, extending urbanization, and the associated forms of transport — severe implications result for the whole of life, implications extending gradually, sometimes not so gradually, into the entire interdependent fabric. Quantitative problems have become qualitative.

Isolated as individuals, daily decisions are made for us, in our name, which effect the course of mammoth rivers, the political composition of entire societies, the nutritional level of continents, and the chemistry of human metabolism — transformations capable of making the entire earth a wasteland or a garden. Decisions are made in our absence about energy production and consumption, agricultural production, urbanization, resource exploitation, the education of children, and the development of our own personality — in most instances without the knowledge, wisdom, or compassion which would allow a proper understanding of our symbiotic role in the one, only family of which all life partakes.

This massive irrationality in the social structure is not simply a case of "low morale," or "lawlessness." It is not a case, as the special environment issue of *Fortune* magazine suggests, that "the wastes that besmirch this land are produced in the course of fulfilling widespread human wants that are in the main reasonable and defensible," or that "we did not get into this mess through such vices as gluttony, but rather through our virtues, our unbalanced and uncoordinated strengths."[4]

iii

The burden of the present work is, of course, to demonstrate quite the contrary. The wastes generated by industrial production do not "besmirch this land," they are killing it. Such waste as mercury, atomic fallout, exhaust from the internal combustion engine, the by-products from the production of chemical and biological weapons, are produced not "in the course of fulfilling widespread human wants that are in the main reasonable and defensible," but as a consequence of a society which, for instance, since World War II has spent over one trillion dollars on "national defense," or war related enterprises. This is not an instance of "unbalanced and uncoordinated strengths," but rather a mirror of the very organization of American society, of the profound inhuman and antiecological bias of America and its European origins. A bias not inevitable, but resulting from conscious choice, and therefore, we must presume, still amenable to transformation.

The appetite with which America consumed the "environmental crisis" in the late 1960's is not unlike the "war on poverty" or the "war on hunger." Except in dimensions, the "war on pollution" assumes all the old familiar trappings — numerous new governmental agencies, a flurry of legislation, and public-directed attention. As with poverty and hunger, the large corporations — responsible for the inequities in income and distribution of wealth — brandished numerous programs to put an end to these social ills. Not surprisingly, the corporations grow fatter and the earth grows leaner. With the environmental campaign, publishers developed a new market overnight. Bookstores instantly acquired row after row of environmental publications, thus further consuming vast amounts of timber and associated resources. Most impressive is the pollution control business which sprung up, promising as much as 20 percent profit for pollution control technology to the same companies most responsible for environmental destruction. In short, as always, America has made out of a social crisis another commodity.

The masters of waste did not overlook any detail in the management of the new commodity. The public acceptance of the term "pollution" is itself a revealing lesson. "Pollution" properly refers to a remediable problem, that is, a problem for which a solution is known and available. When we "pollute" something, says Webster's, we make dirty, impure, or unclean. Thus the emphasis on "cleaning up America," on litter and highway beautification in the Johnson administration. But if we are to take seriously the recent evidence, we must conclude that man now poses a fundamental and perhaps unalterable threat to the evolutionary process. Such a manipulation of the life support system challenges the very regenerative capacity of the earth. The term "pollution" here is completely inappropriate. It harkens back to an earlier, less industrialized, less productive age. It is a measure of the unwillingness to accept the situation at face value that America has chosen to label the condition as "pollution."

In order to comprehend the inadequacy of the pollution concept when applied to present conditions, you might liken the life support system of the earth to the plumbing of your own home. When excessive waste is poured down the kitchen pipes, they become clogged. A vital system which makes the house work no longer functions properly — you have polluted the plumbing of the house. The immediate solution is to purchase a chemical agent which dissolves the clogged mass and allows the water and other material to flow freely again. Usually this is sufficient. At worst, on occasion, we have to replace the pipes. New plumbing is installed. A similar situation exists today in regard to the earth's plumbing, or the life support systems. We have been pouring waste into these systems for decades. The amount and potency have increased steadily. On occasion remedial measures have been applied to repair the damage, such as in the River Thames. The Thames will never be exactly as it was, but much of the nutrient life it once nourished has returned. Today, however, we steadily approach a breaking point in the major life-giving systems — atmosphere, hydrology, nitrogen cycles, photosynthesis. As a result of continued contamination they may well surpass their tolerance level, in other words, break down altogether. The precise tolerance level is unknown. But evidence mounts daily, some presented in this book, that we are precariously close. We will never know precisely the moment at which we begin to recede, or how close we are at any moment. That is because there are as yet no clear and definitive boundaries between the balance of organic and inorganic matter. If the breakdown occurs, it will certainly be more of a whimper than a bang at first. By the time we hear the bang, it will be too late. The earth's plumbing cannot be replaced.

It is today clear that the balance of life on this planet is a tenuous affair. Given estimates as to the potential life of the sun and solar system, it is possible that the biosphere could remain habitable for hundreds of millions of years. Yet, in the Biosphere edition of *Scientific American,* September 1970, G. Evelyn Hutchinson noted that

> many people, however, are concluding on the basis of mounting and reasonably objective evidence that the length of life of the biosphere as an inhabitable region is to be measured in a matter of decades rather than hundreds of millions of years. . . . It would not seem unlikely that we are approaching a crisis that is comparable to the one that occurred when free oxygen began to accumulate in the atmosphere.

This condition does not permit the illusion of despair. The apocalyptic tone of the environmental crusade is itself a major expression of the social order responsible for biological imbalance. The restoration of natural balance depends today upon the destruction of that social order and the birth of a new poetry of human relations. The present task is a candid appraisal of the forces at work — both for death and for life.

Karl Jaspers has remarked that "All life is encompassed within birth and death. But only man knows it." Jaspers was perhaps referring to the unique consciousness man has of his life- and death-wielding power. Second, perhaps most significant for our purposes, the extent to which only man wields such a magnitude of power.

For billions of years the mechanisms which determined what would live and what would die, what species would survive and for how long, were mysteriously bound in the magic of evolution. No animal or species of animal determined consciously the fate of any other, certainly not on a global basis. The complex patterns of diversity and interdependence sought a delicate balance of life and death in accord with available resources. Without conscious knowledge of this process, each animal, plant, or other living creature journeyed its predestined course in ignorance, innocent of the larger drama.

Man stands alone in this regard. Today, for the first time in the history of the earth, a tiny fraction of one species has acquired an awesome power over life and death, over all other forms of life, over the very species of animal which spawned its present existence. This absolutely monumental influence on the course of the planet manifests itself in many guises, both apocalyptic and subliminal. Apocalyptic in the sense of atomic weapons and chemical and biological warfare. Subliminal in the sense that the slow poisoning of the earth is every bit as final. Today a totality of control exists over the forces of death previously unimagined.

This control is not simply relegated to the unhealthy atmospheric conditions of New York or Tokyo, or the possible destruction of Lake Bakail. It has long ago surpassed the regional level and has assumed a global network of manipulation. The lesson to be observed, and observed again, is that man's relationship to nature is no longer determined by nature but is subject to the rule of political management by man. Indeed, not by all men, not by women, but by a finite few.

Politics today has achieved something of an absolute character. While political decision making and control are steadily concentrated in the hands of a very few, the arena of control is steadily expanding. Fewer and fewer people control more and more. This arena of control ranges from the chemistry of the life system to the metabolism of a culture or the personality of its members. It dominates experience, physical and spiritual, separating mind from body, sense from non-sense. The present environmental destruction is increasingly the product of a structure of economic and political power that consolidates and sustains itself through the systematic destruction of the human species and the physical world.

We have failed to comprehend the dimension of this power. While public attention has been focused for some time against racism, war, and militarism — properly so — we have failed to grasp that more people have been killed from the internal combustion engine alone than from all the wars in American history. Just from technical malfunction, in the last seventy years some 1,700,000 Americans have been killed on the highways, while since 1775 some 1,046,000 Americans have been killed in all wars. This of course does not count injuries from the auto, the totals far surpassing those wounded in war. To these figures must be added the vast array of heart and respiratory disease resulting from the internal combustion engine. Emphysema is the fastest rising cause of death in the United States today. This is not to suggest that environmental issues are more important than war, racism, or sexism. Not at all. But it is to suggest that we completely fail to grasp the magnitude of death perpetrated by the American colossus. We must recognize that the suffering and death of war, racism, sexism, and ecological imbalance indeed stem from the same fundamental sources.

While the consequences of atomic war and the new forms of warfare introduced in Indochina are perhaps more apparent and more dramatic, the mechanisms and facets of environmental catastrophe are often less apparent because, in part, we have indeed broken with the naïve evidence of things. Ecological imbalance is today the most fundamental cause of death and disease. Take, for example, cancer. The death rates from cancer in the United States are continually rising, having risen from 18.4 deaths per 100,000 persons in 1950 to 39.1 deaths per 100,000 in 1965, an increase of 113 percent. The carcinogens present in daily environmental contamination are innumerable, those in auto exhausts being only the most highly publicized. The rates of cancer and other (in part or in total) environmentally caused diseases grow geometrically. American medicine still tackles health basically as a case of repair rather than prevention. This is not surprising. To consider health in preventive terms would mean a profound rethinking of the basis of medical and political responsibility. Prevention of cancer today rests as much with the manufacturer of the internal combustion engine as with the medical research teams — perhaps more. It simply is inadequate to seek cures for cancer when what must be done is to eliminate the conditions which spawn the disease — conditions woven into the very pattern of the culture. Such a perspective would involve medical responsibility resting as much if not more with the politician or industrialist as with the physician.

The same perspective must be applied to every facet of our everyday life. The slum landlord literally mortgages the souls and bodies of his tenants. Lead paint peeling off the walls to be eaten by children; inadequate water or heat, cramped conditions, are all fundamental contributors to the health of slum dwellers. The vast white-collar class commuting to and from work — though the stress from commuting is known to affect basic heart conditions; the monotonization of office work in air conditioning instead of fresh air; noise instead of sound, artificial light instead of light — all these aspects of experience are today primary contributors to our physical and spiritual health.

The social imbalance represented by the domination of men over women, whites over blacks, rich over poor, today manifests itself in a variety of disorders ranging from war and race struggle to the breakdown of the nuclear family, the male chauvinist structure of society, and our alienation from our bodies and from all sensual experience. Today the routinization of human experience is at the foundation of our ecological condition.

Ecological balance must therefore be considered simultaneously with the question of political balance. The same irrationality that results in the exploitation of the vast proportion of the world's resources and is responsible for as much as 40 percent of the cumulative pollution of the globe from U.S. domestic sources alone, is a manifestation of the internal concentration of power and control. Of the many thousands of U.S. corporations, the top two hundred control over two-thirds of the manufacturing resources. The automobile and petroleum sectors of the economy, in full array, are perhaps responsible for as much as 40 percent of the Gross National Product. (GNP). Simply put, the political and economic imbalances of capitalism are today inseparable from the chemical imbalances generated amongst the basic elements required for life on this planet.

iv

Already the icons are losing their power over us. The second car in the garage and the lockstep promotion in work gradually lose their attractiveness — considering the physical and emotional cost of achieving them for many. The breakdown of service systems coupled with wildcat strikes is bringing cities to the verge of complete chaos. The ferment throughout America is reflected as much by antiwar protest and black liberation as by consumer concerns over hazardous products, the growing challenge to the nuclear family through communal living, and the movement for the liberation of women.

Unrest reaches into the bedroom as much as into the marketplace, schools, and professions. It is not confined strictly to America. The spread of revolutionary fervor which surged through America in the spring of 1970 after the invasion of Cambodia is tied inextricably to the events of May-June 1968 in France, as well as to the struggle for the liberation of the people of Indochina.

The conditions described in this book speak, as in *A Tale of Two Cities,* of the worst and best of times. It is in this sense that the title *Beyond Repair* is suggestive. It signifies an objective outside simple repair tactics, requiring something beyond. Yet, we are strangers to both worlds. Emigrants from the past, wavering on the threshold of birth and re-creation. These are two glimpses of reality so diametrically opposed as to require a fundamental choice between life and death, birth and stagnation, choice between the power to destroy and the power to create. It is in this sense that questions of our ecological destiny are addressed in this book.

We live today in a world in revolution. America as an empire is but an island. It is a world in which vast numbers of people have determined already to abandon the American destiny of destruction. A world in which the conditions of material poverty can be eliminated, in which spiritual liberation and a truly cooperative society are being born. It is not, tragically, accidental that news of this world is absent from the popular media.

The apocalyptic vision of the environmental crusade is but another facet of the campaign to keep the good news away. We hear nothing but tales of the bomb, the population bomb, indeed the capitalism bomb itself. A world which promises nuclear holocaust, starvation, teeming billions, chemical nightmares, and psychological inertia. We are a product of this daily experience. We act according to the way we see things. Given our destructive experience, little wonder we have such a struggle to create.

The Reality of Constant Threat is a product managed thoroughly by the everyday institutions of our life: home, schools, family, churches, and work place. The world of the apocalypse terrorizes us, isolates us in our fear, forces us into immediate solutions, immediate proposals for fearful, hurried men and women. Never questioning the whole, never sitting down together to plan, to examine carefully, to break out of our social isolation which fear promotes. Never acting collectively except as robots, at the polls or in the streets.

The conditions of our liberation depend upon a promise of a future beyond the present, a space for the new. Any matter-of-fact examination of life on this planet suggests the continual cycle of regeneration, the turning of the seasons, the infinite variety of life and its manifestations. We have not yet shirked our role in this grand scheme. It is for the promise of regeneration, and for the struggle of men and women to achieve it, that this book has been written.

II

From Past to Present

Since man first fashioned a primitive tool, discovered the energy of the sun through fire, developed the first primitive forms of hunting and gathering, the problem of man's place in the world has occupied the human psyche. Today the distance between that first primitive tool and modern weaponry is as great as that between the first primitive cells and Homo sapiens. While man's contemplation of nature is certainly not unique to our present age, man's position in nature is.

As early as the fifth century B.C. man was conscious of himself as a wonder-working force on the earth, a creature with ability to affect the smell of the air, the shape of the land, and the course of waters. Certainly this phenomenon existed long before records indicate. The Greeks, for instance, thought that the object of the investigation of nature was speculative satisfaction. The pre-Socratic philosophers were very much involved in the workings of the natural world and captivated by the wonder and magic of natural things. Heraclitus, for one, understood the profound flux of existence: "Everything flows and nothing abides; everything gives way and nothing stays fixed." Or, perhaps the best known fragment: "You cannot step twice into the same river, for other waters are continually flowing on." The early Greeks understood both the complexity and the hidden harmony which underlies existence. Parmenides speculated that the nature of existence was comprised of four elements: fire, water, earth, and air. Thales recognized water as the basis of all life.

It is with the advent of the Judeo-Christian era that contemporary attitudes toward the world begin to congeal. The contemporary Western psyche remains fixed to attitudes and opinions formulated more than two thousand years ago. The Judeo-Christian view of man's role in the world is derived from the story of the Creation in the first chapter of Genesis:

> And God Said, Let us make Man in our own image, after our own likeness; and let them have dominion over the fish of the sea, and over the fowl of the air, and over the cattle, and over the Earth, and over every creeping thing that creepeth upon the Earth. He shall be fruitful and multiply, and replenish the earth, and subdue it, and have dominion over it. . . .

The universe and all that it comprised existed for man, and for man's sake alone. Utility was enthroned as the first and foremost measure of the natural world and all its inhabitants, human and nonhuman. Pagan animism — the belief that spirits resided in plants, animals, mountains, as well as man — was steadily usurped by the "Spirit of God in Man," and man alone. Other manifestations of the natural world, such as a Sacred Grove known in the East, became a violation of the supreme order of life. Rather than fostering a view of the biosphere which emphasized interdependence and cooperation, this world was valued only insofar as man could elevate himself above natural events and use them for his own advantage. This world was but a way station to another better world. This world, as it were, was to be tolerated. Very early in civilization a hierarchy was established among natural events, those divine, and others. The activity of the natural world was seen much in the same light as the activity of the marketplace; everything as an object, a commodity, to be acquired, owned, bought, and sold. Subjugation of nature was the "divine" instruction.

Ian McHarg has expressed the implications of these origins as well as most:

> Show me a man-oriented society in which it is believed that reality exists only because man can perceive it, that the cosmos is a structure erected to support man on its pinnacle, that man is exclusively divine and given dominion over all things, indeed that God is made in the image of man, and I will predict the nature of its cities and their landscapes.[1]

Monotheism had the "advantage" over more "primitive" world views of providing an orderly, predictable logic to the world; a world of security and purpose that could be passed on from generation to generation, rather than being discovered anew by each.

This view is to be sharply contrasted with the nature-centered Oriental view, in which man has a lesser part. To both we might contrast the more organic, more "primitive" views confronted by the first American colonists at Plymouth: "It is the story of all life that is good to tell, and of us four leggeds sharing in it with the winds of the air and all the green things; for these are children of one mother and their father is one spirit."[2]

Such a world view is shared by people around the world. The Australian aborigine believes that no basic demarcation separates man from nature and life from death. Time is a complete synthesis in which yesterday and tomorrow merge into today.

The basic Western hostility to the landscape and its environs, the hierarchy once established in the natural world, revealed early its analogy in the social order. The precise origins of the idea of conquest — both of the natural and social order — can never be established. However, we can conclude that there exists an intimate relationship between the conquest of nature and the domination of man. The earliest teachings of the Christian era link the conquest of nature with human prosperity. Indeed the struggle for pacification of existence, for mere existence itself, has occupied man from the very origins of civilization. The need for simple necessities: water, food, shelter, were no doubt early inducements to the organization of society along vague notions of the division of labor. Man's control over nature went hand in hand with man's control over other men and women, and with the hierarchical structure of early society developed to provide basic necessities. This is the origin of patriarchy. This structure was sanctioned in the larger environment, in regard to man's privileged position over nature. The struggle for existence no doubt exercised a profound influence on the structure of emerging society; influencing the mode of social fragmentation and hierarchy required to sustain man's domination of the natural world.

The hierarchy of social organization which promoted the Old World explorations — the competition for resources, trade routes, military hegemony — was reflected in the New World treatment both of the landscape and the "native peoples." The first crusades in the Near East manifested much of the savagery and barbarism that was to characterize later explorations, such as those by the conquistadors in Mexico and Peru and the colonists in North America. The very act of exploration and discovery reveals at the outset the implications of the hierarchical structure. The landscape and people of the North American continent did not exist in the eyes of the explorers until they were "discovered."

The man-dominated world view was predated by another, perhaps equally influential conception of hierarchy. Lewis Mumford[3] has documented the origins and rise of the Megamachine, culminating in his latest work, *The Pentagon of Power*. The power nexus which originated as early as ancient Egypt with the construction of the pyramids, together with the Judeo-Christian man-centered world, found its ultimate expression in the middle of the sixteenth century, when "the mechanical world picture, was first conceived as the expression of a new religion and the basis of a new power system." Mechanization existed well before the seventeenth century and the later "industrial revolution." Together with absolutism in government and the organization of the capitalist economy, the mechanistic advances brought with them an emergent world view in which the diversified and balanced system of medieval guild protection and polytechnics was usurped by the monotonization of mass production and the consolidation of ownership under the banner of laissez-faire promoted by Adam Smith.

With the coming of the bourgeois revolution, instinct and tradition were replaced by taste. Mechanical regularity replaced ritual regularity. Large scale mechanization brought with it military discipline and the cash nexus, and, Mumford notes, that "impersonal authority and submissive obedience, mechanical regimentation and the control of humanity went hand in hand."

Already clearly evident in the Renaissance, cemented in the seventeenth century within the emerging world view, was the notion of progress. Put in the most simple formulation, "things would get better, every day, in every way." The built-in implications entailed a positive faith in the future. If events should not get better here on earth, there was always a better place to go. Salvation thus came to rest increasingly upon faith rather than action. The political implications of this ideological transformation were to have profound consequences for liberal democracy. Progress was a deeply static concept, expressing confidence in the existing forms and institutions. Faith in the established forms and good intentions would evoke the necessary and required changes that would move the populace increasingly toward a better condition. The mystique behind the Copernican revolution lent authority and conviction to all social manifestations of order, regularity, and predictability. The centralization of power and control in the emerging leviathan was analogous to the sun's own position and influence in the galaxy.

The new science of progress ushered in the spell that was cast upon the New World strivings. The initial discoveries in the astronomical laboratory are today reflected in the computer-automated, remote-control assembly line.

The orderliness of the new science, profoundly utilitarian in form, was bound inextricably to the Judeo-Christian notions of human dominance and power over the natural world. The products of science assumed a similar value to those of the production line. Both were employed first and foremost to achieve control, regularity, order, and domination over everything irregular or unorderly.

It was not until Descartes, and more importantly, Bacon, that science became firmly employed in the service of power. Theoretical science gave way to practical science. Science became the tool par excellence in the historic struggle for man's conquest of nature. In this sense we might suggest that science, as with technology, is never a neutral element. It arises, is conceived, and is employed always in a particular historical context. Referring to the use to which the new Cartesian method would be put, Descartes exclaimed, "To render ourselves the masters and possessors of nature . . . to enjoy without any trouble the fruits of the earth and the good things which are to be found there."

The most illuminating document of the new science was Bacon's *The New Atlantis*. Bacon is clear about his intent: "Our main object is to make nature serve the business and conveniences of man." Or, from his *The Masculine Birth of Time:* "To stretch the deplorably narrow limits of man's dominion over the Universe to their promised bound." It is evident from Descartes and Bacon that the predominant attitude of the time, one not unchanged in our present era, was that the steady and regular expansion of the human conquest of nature, through science, would yield new and increasing benefits for mankind. The highly pragmatic tone to the adventure without limitation or reservation translated scientific invention into technological innovation, innovations which were destined to greatly extend man's capability in the constant struggle for existence. Progress was thus inevitable.

With those who first set out from the Old World to the New came a firmly established orthodoxy about man's role in the natural world, indeed his duty. The steady separation and elevation of man apart from nature was to deeply determine the form of the experiment in the New World. Indeed, Mumford concludes that the "new world man had dug his own grave before he was out of the cradle." He reports one observer as noting that within six years after Columbus set foot in America, the Spaniards had killed over half a million Indians.

ii

This was in large measure the heritage that John Winthrop brought to Plymouth Rock. The construction of the "New Jerusalem" in a "hostile land," the move westward, the elimination of the Indians, buffalo, Mexicans, and the very landscape itself, the complete assumption of control over the natural forces was shouldered as the special mission of the American people. The wilderness was the fundamental conditioning factor of the American genesis. Success as a pioneer was measured by one's ability to conquer the land. Very early in our history we constructed a mystique about nature, as something to be camped in, hunted, or hiked. As something to be used.

The Puritans ordained a strict utilitarian standard for nature. The order they wished to establish in the face of the wilderness was applied with equal vigor to social organization. The deep sense of righteousness and conformity within the Puritan covenant injected a heavy dose of morality into the confrontation with the natural world and the challenge of building a "New Jerusalem." Purity rather than democracy was the early goal.

As settlers in a New World, they were colonists. On the one hand they rejected the old order of injustice, intolerance, and servitude. On the other hand, they were quick to establish their own "second-class citizens." The Indians were driven from their homes, their land, and finally, what was left of them, into reservations, today known as concentration camps. Indians were not the only class excluded from the new covenant. Women were little short of slaves. They owned no property and thus were not included in the activity of government.

It is not accidental that nature itself was ascribed the feminine gender. Women, as nature, were mysterious, dangerous, volatile, and therefore something to be conquered, indeed, to be raped. The hierarchy of men over women ran extremely deep into the fabric of society. It affected the activity of the marketplace as well as the home. Women were considered chattels, to be acquired by a man. The customary marriage ceremony reads, "man and wife," not man and woman.

While America early espoused a rhetoric of inclusion, in reality it established one of exclusion. James Madison was the primary architect of the new social order, "Societies are divided into different interests and factions," in accord with all manner of distinction: finance, political persuasion, religion, geographical distribution, et cetera. Madison believed that "government is instituted to protect property of every sort, as that which lies in the various rights of individuals. This being the end of government, that alone is a just government, which impartially secures to every man, whatever is his own." Thus the purpose of government is to secure for every man that which is his *own.* A man's interest is defined as that which increases his *individual* welfare. Politics is relegated to a very narrow aspect of one's life. Those qualities which seem most important to life — love, affection, friendship — are defined out of the political sphere. What men have in common is thus simply a place where people barter for scarce goods. Citizenship is relegated to castlemanship, each to his own. The ruled in no way partake in the ruling.

The result of the early Constitution was to generate a profound sense of anxiety, isolation, and pointlessness in the political quality of life. The sense of limitlessness in the social context was not without parallel in the physical dimension. The wilderness engendered a sense of open spaces, of space without limits. The conquests of Davy Crockett and Daniel Boone attest to the continual struggle to subdue and conquer the unknown virgin territory. The very fabrication of the early four-walled forts symbolizes the establishment of civilization amidst the wilds. Whether in the struggle against matter or mind, we witness the continual fragmentation of experience and outlook, thwarting always the kinds of synthesis which might reconcile the political with the biological.

This is not to suggest that there have not been those Americans who understood better the conflicts between "nature and civilization." During the later decades of the nineteenth century the voices of Emerson, Melville, and Whitman stand clarion. Later, in 1904, Henry Adams forecast well the future of this driving power-complex behind man's domination of the natural environment. "Prosperity never before imagined, power never wielded by man, speed never reached by anything but a meteor, had made the world nervous, querulous, unreasonable, afraid." In another passage he suggests that "law . . . would disappear as a theory and give way to force. Morality would become police. Explosives would reach cosmic violence. Disintegration would overcome integration."

In the middle of the nineteenth century a Frenchman, Alexis de Tocqueville, captured prophetically the image of the future. In *Democracy in America* de Tocqueville systematically described "the sort of despotism that democratic nations have to fear." He chronicled the already evident separations among men, the triumph of the commercial method within the political sphere, the position of blacks, the dread of fundamental change, and, perhaps most prophetic, the love of power: "Democratic nations often hate those in whose hands the central power is vested, but they always love that power itself." He was, as well, aware of the human hierarchy at work: "I am persuaded that in the end democracy diverts the imagination from all that is external to man and fixes it on man alone. Democratic nations may amuse themselves for a while with considering the productions of nature, but they are excited in reality only by a survey of themselves."

There is one astonishing passage in de Tocqueville which summarizes the nature of democratic despotism in a fashion yet to be matched. The following was written some hundred and twenty years ago:

> ... As for his fellow citizens, he is close to them, but he does not see them, he touches them, but he does not feel them, he exists only in himself and for himself alone....
>
> ... the supreme power... covers the surface of society with a network of small complicated rules, minute and uniform, through which the most original minds and most energetic characters cannot penetrate, to rise above the crowd. The will of man is not shattered, but softened, bent, and guided; men are seldom forced by it to act, but they are constantly refrained from acting. Such a power does not destroy but it prevents existence....
>
> I have always thought that servitude of that regular, quiet, and gentle kind which I have just described might be combined more easily than is commonly believed with some of the outward forms of freedom, and it might even establish itself under the wing of the sovereignty of the people.[4]

Such outward forms of freedom as de Tocqueville suggested have today been documented well, such as the role of "tolerance" in a "pluralist democracy," in *A Critique of Pure Tolerance,* by Wolff, Moore, and Marcuse.

It did not, of course, require a Frenchman to observe what was happening. In 1837 Emerson wrote, "The State of society is one in which the members have suffered amputation from the trunk, and strut about so many walking monsters."

In these later years of the nineteenth century fundamental choices were steadily being made which were to shape profoundly the course of American development. In little more than one hundred years America would indeed reach the very last frontier, the domination of the unknown — the challenge to the evolutionary process itself. I have quoted in this chapter from this period to demonstrate that what we witness today in both the ecological and social crises did not occur overnight, but had its fundamental origins in the very growth and ascendancy of America as a world power. These were years of decision. It took the American leviathan until 1830 to reach a one billion dollar GNP. By 1880, a mere fifty years later, it had swelled to ten billion dollars. Involved in this staggering economic development was a process more profound in its import for the future of America than all the early debates between naturalists and industrialists, more than the struggles of John Muir or Whitman. Here, finally, the gleam in the eye of John Winthrop had begun to sparkle, at long last we had "built a city upon the hill," and "the eyes of the world were upon us."

The outline of the American dilemma was by now quite clear: in the scurry to tame the bountiful wilderness and provide for the immediate necessities of civilization, an edifice had been begun already with the traits which later were full blown to threaten the very mission itself.

The apocalypse is not an invention of Hiroshima or the current environmental crusade. As early as Jonathan Edwards, the day of fire and brimstone was very real in the American psyche. Indeed, one is tempted to suggest that it is nothing less than the propelling force in the American conquest of nature. Perry Miller has chosen the theme of the apocalypse to close a series of unique essays entitled *Errand into the Wilderness*. "Humanity lusts after the apocalypse, even after nature seems unlikely to provide it. The human finger actually itches for the trigger. . . . Not for this was the errand run into the wilderness, and not for this will it be run. Catastrophe, by and for itself, is not enough."[5]

In innumerable ways the initial acts of this great American drama foretold those of the present — the plight of blacks and other minorities, the role and status of women in America, the domination over nature, and the inequities in income and wealth. An examination of both the spirit and effect of the early conservation movement during the Progressive Era reveals the die which later was to cast the present ecological crisis. The conservation movement began with the works of men such as Thoreau, Emerson, and Muir. But it was not until the time of Theodore Roosevelt, and more importantly, Gifford Pinchot, during the Progressive Era, that a number of limited measures were begun under the vague and all-encompassing rubric of conservation.

iii

The term "conservation" was first applied in America to the movement to construct reservoirs to conserve spring flood waters for use later in the dry season. In his definitive study, *Conservation and the Gospel of Efficiency,* Samuel P. Hays describes in detail "the concept of planned and efficient progress which lay at the heart of the conservation idea."

From the problems of water management the conservation idea came to grapple with forest use, grazing lands, and virtually the whole range of resource and development issues. At the height of the conservation flurry, after 1908, the term became "highly elastic," in Hays's words, to include almost everything that a wide variety of reformers ascribed to it, much as in the present day. While Roosevelt received the bulk of public acclaim, he did little more than lend his prestige and support to those who generated the movement, foremost among whom was Gifford Pinchot. In *The Fight for Conservation,* Pinchot spells out his philosophy of development: "The welfare of this generation first, and afterward the welfare of future generations to follow"; of preservation, "The first duty of the human race is to control the Earth it lives upon"; and of the common good, "natural resources must be preserved and developed for the benefit of the many, not merely for the profit of a few."

In short, the essence of the conservation idea was planning for the efficient development and use of all natural resources, as opposed to a robber baron, spoil mentality. Hays documents how the men around Pinchot "became the nucleus of a group of federal scientists and technicians whose search for greater efficiency in economic growth gradually committed the administration of Theodore Roosevelt to a wide program of natural resource conservation." This movement "emphasized expansion, not retrenchment; possibilities, not limitations." Thus we must realize that the legislation issued in the Progressive Era was not motivated by a questioning of the distribution or ownership of wealth and resources, but by a question of method: the problem of finding the most reasonable technique to promote efficient growth on the part of those who already controlled land and resource patterns. The conservation movement in fact was built around the difficulties of management, rather than ecological diversity and stability.

Perhaps the most significant division in the early conservation movement was that between Gifford Pinchot, and John Muir, who founded the Sierra Club. This dispute forms the threads in the present-day fabric of the conservation movement. In the battle to preserve the Hetch Hetchy Valley Yosemite from flood control development, John Muir desired to preserve the "pristine wilderness in its original glory." Pinchot argued for the "wise use" of natural resources and the guided development of Hetch Hetchy. The battle became a cause celèbre. The position of Gifford Pinchot finally triumphed and continues to dominate present land use patterns. However, both positions, from an ecological and political perspective, are inadequate. On the one hand, nothing in nature is static, nothing can be truly preserved. The constant flux of the natural process denies us the knowledge of original creation. Muir did not anticipate the extensive use of technology and the global spread of chemical poisons. He did not foresee, as did John Adams and many other predecessors, the escalation of power and destruction in the American grain. The static world view, in which anything is permitted to remain, unchanged, is completely undermined by modern relativity theory, and present understandings of ecosystems. Even Heraclitus, thousands of years ago, was aware of the constant flux. On the other hand, to assume that natural resources exist for man's use is to fall victim to the arrogance and hierarchy that have for ages dominated Western man's view of the world. The historical context of the notion of "utility" is bound to an arrogant relationship with nature. The Greeks called it *hubris.*

Neither Muir nor Pinchot properly understood the natural processes. Hans Selye has described the symbiotic relationships of an organism as "intercellular altruism." He meant by this that in a complex organism the varied cells each contribute to the functioning of the organism, and in return the organism responds to cellular processes. This constant give and take is symbiosis, in which a reciprocal relationship is established with benefits and responsibilities to all partners. Utility is quite a different relationship, involving a hierarchical structure which functions as domination. Life is completely dependent upon flux. Neither static nor utilitarian notions are compatible with our understandings of how the life processes function.

The realization today that natural resources are considered in terms of use — the fee we pay to enter the park, which means paying someone else to provide the care and reciprocal process we should have provided — demonstrates how distant our cultural norms are from an involvement in natural systems on an interdependent and organic basis.

The function of the Pinchot ideology was revealed by Hays: "The broader significance of the conservation movement stemmed from the role it played in the transformation of a decentralized, nontechnical, loosely organized society.... into a highly organized, technical, and centrally planned and directed social organization which could meet a complex world with efficiency and purpose."[6]

Pinchot and Roosevelt constantly emphasized the value of large-scale business organization. What was desired was to curb the activity of small speculative fly-by-night operations, in order that larger corporations with more capital and resources, could provide for a more stable and rational development over a longer period. Development, of course, was geared to the maximization of growth and profit on the part of the corporate managers.

Hays further notes that

> the conservation movement did not involve a reaction against large-scale corporate business, but, in fact, shared its views in mutual revulsion against unrestrained competition and undirected economic growth. Both groups placed a premium on large-scale capital organization, technology, and industry-wide cooperation and planning to abolish the uncertainties and waste of competitive resource use.

Thus the regulatory agencies were born. The most obvious answer to the above concern was to establish one single resource authority which could manage the natural environment in the most efficient and profitable manner. So we find that, contrary to popular belief, it was industry itself that in most cases promoted the establishment of regulatory agencies. Gifford Pinchot worked hand in hand with private landowners and Federal foresters in drawing up management plans for the land. In the case of grazing lands, large industrial corporations worked hand in hand with the government. Present-day conservation organizations work in the same manner.

Thus, as Gabriel Kolko demonstrates brilliantly in *The Triumph of Conservatism,* monopoly was not the accidental happening it is taken to be, but the result of very concrete relationships established during the Progressive Era between politics and economics.

> It is the business control over politics rather than political regulation of the economy that is the significant phenomena of the Progressive Era. Such domination was direct and indirect, but significant only in so far as it provided a means for achieving a greater end — political capitalism. Political capitalism is the utilization of political outlets to attain conditions of stability, predictability and security — to attain rationalization in the economy.[7]

It is for this reason that crucial big business support can be found for virtually every regulatory agency established. When one looks at their origins, it is not difficult to understand the reason why every regulatory agency has become the virtual captive of the industry it regulates.

National regulation of industry works to the clear advantage of maintaining the national character of industry. In other words, Federal regulation establishes the national character of the commodity market, stabilizes it, and tends to regularize the allocation of resources and capital. Moreover, national regulation tends to unify the national nature of production along with the nationalization of consumption. Particularly in the case of consumer regulation, such regulation tends to legitimize and standardize product distribution, through such measures as "truth in lending, " "truth in packaging," et cetera. In many cases, industry gets the government to provide the capital that extensive product testing might require. The general tendency of industrial regulation inevitably leads to either monopoly or oligopoly, toward a growing standardization and stabilization of the market. In this sense, the conservation movement of the early years of this century was part of a larger, more fundamental event, the economic consolidation of private capital through public means — Federal regulation!

The movement for Federal regulation during the Progressive Era was destined to have a far-reaching effect upon direct participation in government. The increasing reliance upon expert management and efficient control — the hallmarks of early conservation practice — gradually and steadily excluded the general public from either knowledge or involvement in resource allocations. This, in turn, generated a class of scientific managers provided by industry to work in governmental positions. Thus, today, environmental issues are hotly debated among scientists while the general public is excluded from comprehension or participation.

There can be little doubt that such a development was preferable to big business. A popularly based social movement might have raised more fundamental questions than use and management, such as private property or the ownership of wealth and resources. Federal regulation, moreover, freed the Justice Department from the necessity of considering more basic questions such as economic concentration. Gradually, private capital and public capital began to merge during the Progressive Era. Government became increasingly involved in the subsidy of such activities as the testing of products, safety monitoring, research and development — all activities paid for by the general taxpaying public, but which profited only private parties. This consistently contributed to the organization of the society around social lines for private gain.

Conservationists who rallied behind business in its drive for self-regulation — increasing dependence upon expertise, large capital, and the nationalization and stabilization of the commodity market — were laying the foundations, however unwittingly, for the eventual destruction of the very land they so cherished. Were it not for the massive consolidation and expansion of the American economy during the Progressive Era, the imbalance between the structure of society and the natural life support system would not have achieved the ultimate contradiction it now manifests. With the increasing concentration of the ownership of capital and natural resources came the increasing ownership of the air, land, and water. The last fifty years have witnessed the continual expansion of the forces of production and consumption, until the mere dimensions challenge the life support system's capability to sustain them.

James O'Connor has pointed out that as the "twentieth century wore on the owners of corporate capital generated the financial ability, learned the organizational skills, and developed the ideas necessary for their self-regulation as a class."[8] Federal regulation thus, more than anything else, revealed the class nature of American society, for it virtually assured that the process of production would remain in the control of the class destined to directly benefit.

Above all, the sheer scale and concentration of economic power — mostly developed during the Progressive Era — is today responsible for the wide-scale destruction of our environment. It is only with massive productive capabilities, international marketing patterns, products so many and so numerous as to preclude adequate testing, that the global destruction resulting from atmospheric contaminants, solid waste, and chemical poisoning becomes possible. The conservation activity of the Progressive Era served thus to further enforce the utilitarian, mechanistic, and exploitative attitudes of man over nature. It cemented the role of science as power, and scientist as expert in the determination of resource allocations. Scientific management was the goal, industry-prompted Federal regulation the tool.

It was assumed that government would provide public protection for those needs not reflected in the marketplace. In other words, government would balance the competing interests of public welfare and industrial profit. Is there anyone today who doubts which way the scales have tipped? The interest politics of James Madison found its suitable abode in the partnership struck between government and industry during the Progressive Era. The regulatory agencies were established, staffed, and supported at the outset by the industry they were to regulate; this is as it was cast by Madison. The more propertied the interest, the more legitimate its right to rule.

By the end of the Progressive Era conservation became a squabble among the ruling classes, a case of proper management, rather than addressing the dialectic which saw the imbalance man generated over nature as but the other side of the same coin as the emerging imbalances among men and women in regard to wealth and participation in the decision-making process of society. But conservation, from the outset, was never concerned with balance, but with progress. The faint voices of opposition, such as Muir, were never really heeded.

Conservation became, and remains, an elite club. The term grew to become "highly elastic," encompassing a variety of upper class causes which relied upon "moral tonic" rather than the objective conditions of oppression as the basis of the movement. The mild-mannered nature of the conservation crusade today can largely be attributed to the moral rather than objective motivation of middle class whites. The texture and attitude of the Progressive Era survive and flourish among modern day conservationists. It is perhaps the highly moralistic and highly individualistic tenor of the movement which most links present to past — perpetuating the very mentality which is the greatest obstacle to ecological and social balance.

iv

Samuel Hays's assessment of the conservation crusade of the early twentieth century is, with a few alterations, an adequate description of the present ecology fervor. Hays suggests that the momentum of the early conservationists was in fact a reflection of something deeper, a feeling on the part of

> the American middle class which shrank in fear from the profound social changes being wrought by the technological age. These people looked backward to individualist agrarian ideals, yet they proposed social planning. . . . A vigorous and purposeful government became the vehicle by which ideals derived from an individualistic society became adjusted to a new collective age.[9]

Now as then, the American middle class feels the deep anxiety resulting from profound social and technological transformations they neither understand nor control. Middle class housewives cringe at virtually every product bought in the supermarket for fear of some new chemical discovered with dangerous consequences. The cities are a technological horror house — stalled subways, power failures, machines too complex to understand and too hip to abandon. Middle class people no longer burdened with the requirements of income and food, feel more deeply the more subliminal, more affluent forms of oppression which blacks, poor people, and the unemployed do not have the luxury to consider. The clamor after Cambodia in the spring of 1970 for a new Congress, a more aggressive conservation ticket, and the flurry of conservation legislation — all are reminiscent of the Progressive Era. Then, as now, the urging that industry become involved, that they assume a more "responsible" involvement in the ecology crusade. Thus, the flurry in the pollution control technology represents an incentive to industry today much in the same manner as Federal regulation was an incentive to industry during the Progressive Era — a means for the further consolidation of capital and market position by the very sources most responsible for the ecological devastation.

Modern-day conservation organizations reflect without exception the impoverished heritage of the Progressive Era. Organizations such as the Sierra Club, the Wilderness Society, the Nature Conservancy, Friends of the Earth, Zero Population Growth, and the host of others, always view problems in the defensive, seeking most often to repair damage already done, rather than eliminating the source of that threat. Their actions are fragmented, attacking isolated problems and situations rather than complex patterns of social and political behavior which are the source of all particulars. Within the last decade these organizations have secured four national parks, eight recreation areas, and some one hundred wilderness preserves. That large areas of semiwilderness should be protected from development is without dispute. However, this is a small measure of the threat involved.

A wilderness area might become a park, preventing economic development, without its being realized that the area is at the same time threatened with chemical destruction, the foliage and animal life decimated by the spread of chemicals such as DDT and mercury, as well as automobile emissions. The rapid deterioration of the forests in Southern California is an example, which is typical of many areas adjacent to large industrial towns and cities. These are but the more obvious examples; the same process is occurring in the wilds of Canada and throughout the world. Within the century man has eliminated 90 percent of all African wildlife. Present conservation measures are in reality only stopgaps, always provisional. The insistence upon "preserving" wildlife areas without a simultaneous understanding of and struggle against the less obvious — but in the long run more hazardous — chemical threat, reflects the old heritage of static understandings and conservative mentalities.

The internal structure of conservation organizations is revealing. They are without exception white, affluent, and professional in orientation, resembling something more of an elite outdoor club than a movement for social change. That is because in part only the affluent can afford to "take advantage" of wilderness areas. The poor and almost poor do not have the "time off," nor the methods of transportation required. At a recent Southwestern Regional meeting of the Sierra Club, a speaker from the Oil, Chemical, and Atomic Workers talked about the hazards in the industrial working place. The audience was not responsive. Afterwards, he was told that "workers are our enemies, not our allies." This attitude is quite typical of the basic professional, upper class white orientation of all conservation activity.

The class orientation of the conservation movement was nowhere better demonstrated than in the Earth Day activities of April 1970.[10] At the same time that student participation in the struggles for peace and black liberation was leading students off the campus and into the streets, the Earth Day teach-ins proposed that students should return to their campus and engage in orderly, rational dialogue with industry.

In January 1970 the national editor of the Hearst newspaper group circulated an article that was widely reprinted, entitled, "The Issue That Could Unite the Generations." It said, "We cannot stop constructing highways, airports, buildings, or housing, but such projects will have to be developed within a new context of environmental thinking." Nothing in this context would be more desirable than a consensus which unified people by their relation to commodities such as the Hearst Press rather than their authentic needs and desires. Fundamental social transformation of any society has never been achieved through consensus, but rather through struggle — struggle which separates those who rule from those who are ruled. Polarization in an oppressive society is the precondition for liberation.

Current conservation organizations have scant understanding of the governments of oil, chemicals, agri business, and the like. They fail to grasp that it is man and the political-economic structure he supports which determines the natural fate of the earth — not the rivers, streams, and mountains themselves. The constant insistence upon "everyone being in the same boat" reflects a profound ignorance of the origins of our struggle — origins which are intimately woven into the social structure of our society, not the biological chains which tie people together.

The suggestion by the environmentalist that "environmental issues transcend other issues," and thus we should "support a man from any political party if he is a true Friend of the Earth," must be seen for what it is — another method of sustaining the privileged position of the conservationist who shrinks from questioning fundamental assumptions about the basis of political power, production, and the very viability of the American dream. Many conservationists in the state of Washington supported the reelection of Henry (Scoop) Jackson because he had assisted them in obtaining several national parks, while at the same time Senator Jackson has been a leading hawk in the administration's war policy, as well as the prime defender of strategic air attack and nuclear parity. Is Senator Jackson a Friend of the Earth?

Conservation organizations must realize that it is precisely the separation between views regarding the war and views regarding a national park which is at the root of our dilemma. What is required is a holistic understanding of the sources of domination and control, an understanding motivated by the necessity to transform social structure, not merely preserve wildlife areas. The elitist composition of conservation organizations must move from the publication of handsome twenty-five-dollar books to true participation in the struggles of other oppressed elements of the society. Such a movement on their part will require fundamental internal struggle, like that entailed by every true liberation movement.

The half century of conservation lobbying finds its just rewards in the programs they have fought so hard to attain. Before the inauguration of Richard Nixon, Daniel Patrick Moynihan circulated a now famous memorandum in which he urged the new President to exercise "benign neglect" in regard to the issues of race in the United States. This policy is hardly an innovation. Successive Presidents have exercised "benign neglect" regarding the issues of the environment for some time, to be matched in fervor only by the Congress. While the clamor for increased Federal action swelled in the closing years of the 1960's, Federal expenditures as a percentage of the national budget actually *decreased.* From 2.3 percent of the budget spent on all natural resource programs in 1965, the proportion had dipped to 1.8 percent by 1970.

1965	2.3%
1966	2.2%
1967	2.0%
1968	1.9%
1969	1.9%
1970	1.8% (estimated)[11]

SOURCE: *CF Letter,* published monthly by the Conservation Foundation, Washington, D.C., March 17, 1969.

Even these figures are deceptive, for they reflect the authorized expenditure, not what in fact was appropriated. An examination of the gap between authorization and appropriation in regard to air pollution control and water treatment plants alone is illuminating. See p. 34.

These figures should not come as a surprise. As Dr. Lee DuBridge, former science advisor to Richard Nixon stated in 1970, "Let's face it, waste products are a fact of life we have to recognize. The more active, the more industrial a civilization we have, the more waste products we are going to produce. Clearly the U.S. will be producing more waste in the future — not less." In other words, the annual discard of 7 million cars, 100 million tires, 20 million tons of paper, 28 billion bottles, and 48 billion cans is just the beginning!

(in millions)

Air Pollution Control	*1968*	*1969*	*1970*	*Totals 1966-1970*
Authorization	$109.0	$185.0	$134.3	$504.8
Appropriation	64.2	88.7	95.8	315.4
The Gap	44.8	96.3	38.5	189.4
Treatment Plant Grants				
Authorization	$450.0	$700.0	$1,000.0	$2,550.0
Appropriation	203.0	214.0	214.0	1,038.0
The Gap	274.0	486.0	786.0	1,512.0

SOURCE: Conservation Foundation [see source in above chart].

We must understand that it is not simply that not enough funds were spent on Federal programs, or that they were ill-executed. For what Federal "regulation" was intended to regulate was not so much the emission of effluents or the production of waste but the political climate in which such crimes are permitted to occur. The intention of Federal programs has never been the elimination of hazardous emissions, but to regulate their flow, to monitor the emission. It is a matter of management. The difference between seeking to eradicate hazardous agents from the environment and simply wanting to regulate their production and distribution is the crux of the issue. One position assumes the legitimacy of production at the outset, if merely "regulated"; the other position denies the legitimate right of an industry to contaminate the life support system for any reason. Today, hazardous emissions are permitted if the polluter is willing to pay a fine — in most cases a few thousand dollars charged to a company whose net profit reaches into six, seven, or eight figures. As *Forbes* magazine put it, there is "cash in all the trash."

The virtual impossibility of obtaining adequate information about actual emissions is illuminating. On October 21, 1970, the Federal Water Quality Administration initiated a national industrial waste inventory, something the House Committee on Government Operations had continually requested since 1963. The committee reports how it has "patiently and persistently, but until recently unsuccessfully," sought to convince Federal executives about the need to develop information about the type and extent of industrial waste discharge. The information may take several more years to accumulate, if it is ever done. How can the cost of these years be measured for plants, rivers, the atmosphere, and our own chemistry?

This is not an isolated example of what is involved. Section 102 of the Environmental Policy Act of 1969 requires that an environmental impact study be made prior to environmental policy decisions. In November 1970 the Nixon administration decided it could withhold such studies from the public until such time as the decision was made; as Russell R. Train, chairman of the President's Council on Environmental Quality stated, "There is no question that there is a tendency to prepare a 'Section 102' environmental impact statement after a decision has been made." As such, the importance of the act was suggested by Train: "The public gets a retrospective look, and their impact comes largely as some comment . . . about the same decision in the future."

At present the public has virtually no information about the scale or extent of chemical emissions on the part of industry. One illusory step taken to remedy this situation was the announcement by President Nixon on December 23, 1970, under the 1899 Refuse Act, that an estimated 40,000 industries will be required in the future to obtain permission to dump waste into Federal waterways. While the most extensive of present efforts, this still does not apply to ocean dumping, to municipal sewage plants, agricultural and real estate development projects, or two-thirds of all emissions into waterways. At present, less than a third of the 17 trillion gallons of industrial waste is treated even once. The impact of the proposal would be a future licensing apparatus to dump wastes. Yet,

Democratic Senator Joseph Tydings of Maryland, among others, has documented how numerous government agencies notoriously charge fees to industry so far below their actual administrative cost that taxpayers have to foot the bill. An example mentioned was the Interstate Commerce Commission which charges a five-dollar renewal fee for carriers transporting explosives even though processing each application costs the commission four hundred and thirty dollars.

Such relations between government and industry are of course nothing new. On April 9, 1970, President Nixon established by Executive order, the National Industrial Pollution Control Council with sixty-three officials from major industries. The administration requested $475,000 for operations for the leaders of industry, nearly one-third the total amount requested for the Council on Environmental Quality. The President was merely providing a public face for a relationship which has existed privately for some time, beginning in the Progressive Era. Such government-industrial intimacy goes a long way toward explaining a consent decree signed on October 28, 1969, to a civil antitrust suit brought by the Justice Department on January 10, 1969, against General Motors, Ford Motor Company, Chrysler Corporation, American Motors Corporation, and the Automobile Manufacturers Association charging them with illegally conspiring to eliminate all competition among themselves in research, development, manufacturing, and installation of motor vehicle pollution equipment. The charge also alleged that the industry had eliminated competition in the purchase of patents and patent rights from other parties covering motor vehicle pollution control equipment from 1953 to 1969. The consent decree, negotiated by Detroit with the Nixon administration merely amounted to an admission that Detroit would cease further stalling on the development and installation of antipollution devices. No dates were established. Other such instances are innumerable. In January 1971, a Federal prosecutor who sought to enjoin General Motors from polluting the Hudson was dismissed, without explanation.

Even if ameliorating regulatory legislation could be obtained, free of industrial pressure, the task of enforcing any regulation applied to 40,000 different industries — each with many facilities —

is impossible. This can readily be demonstrated if one considers the success of Federal agencies in simply monitoring the quality of municipal water supplies. On August 17, 1970, the Department of Health, Education, and Welfare issued a report entitled, *Community Water Supply.*[11] The study examined 969 representative water supply systems in eight major metropolitan areas which service 18 million people. The study revealed that several million Americans were drinking "potentially hazardous water." While the findings indicated that only "2 percent of the people surveyed (360,000) were being served by systems whose average water quality was potentially hazardous," the almost complete lack of control or adequate supervision of the water quality of practically all the systems surveyed points to the incapability of the government to do such a job.

> ITEM: . . . Seventy-nine percent of the systems had not been inspected in 1968, the year prior to the survey, by state or county officials. In 50 percent of the cases, plant officials did not remember when a state or local health official had last surveyed the supply.
>
> ITEM: Frequent sampling of drinking water to assure the absence of bacterial contamination is essential to the operation of any water supply system. Ninety percent of all systems failed to meet the survey's criteria for frequency of sampling during the twelve months prior to inspection.
>
> ITEM: Seventy-seven percent of the plant operators were inadequately trained in fundamental water microbiology, and 46 percent were deficient in chemistry relating to plant operation. . . .

Only twenty-five potentially hazardous substances were checked for. And yet there are over twelve thousand toxic chemical compounds in industrial industrial use today with another five hundred developed each year — most of which find their way into water supplies. The proliferation of chemicals in use — weed killers, fungicides, pesticides, phosphates, trace metals such as mercury, lead, cadmium, acids from mine drainage, antibiotics and hormones, as well as many agents which are potentially carcinogenic becomes more threatening every day.

In this context, supervision becomes ludicrous. Altogether, there are hundreds of thousands of chemicals in daily use which find their way into the land, air, water, and our bodies. There simply is no adequate means, after production, to test for such a number of agents, or to do the incredibly extensive monitoring that would be involved. The 40,000 outlets mentioned above are a very small fraction of the total site releases for hazardous agents. The scale of industrial production and distribution defies human supervision. The prohibition comes not so much from the lack of technical means — though that itself is a task to dwarf in magnitude the moon explorations — but from the political climate which assumes at the outset the fundamental right to pollute if one's pocketbook can afford the fine. This political climate results from a structure of commodity production beyond public control or influence. Nothing short of a radical decentralization of production would make any difference in regard to the production of chemicals. Such a decentralization would tailor the local production of chemicals to meet designated public needs — designated not by the marketplace but through some form of democratic process. In other words, nothing short of the community control of resources would be adequate to bring about a balance between the production of chemicals and the environment. This is a direction which will be discussed at greater length toward the end of the book.

The Federal environmental programs will rely increasingly upon tax mechanisms as both the carrot and the stick in the "war on pollution." Aside from money raising, taxation, it is argued, can also encourage or discourage certain kinds of behavior on the part of industry or the consumer. Only recently has the Federal debate picked up on devices employed by the states since the mid-sixties; tax incentives to industry to install pollution control equipment.

Tax incentives are generally of three forms: property tax abatements (available presently in twenty-one states); sales and use tax exemptions (thirteen states); and as in income tax an accelerated write-off or special tax credit (eight and five states respectively). On a Federal level, the administration's proposed tax on the leaded content of gasoline was abated at the first opportunity in the House Ways and Means Committee. At the time of writing, the administration has proposed a penalty tax on the sulphur content of coal, oil, and natural gas. The proceeds, unlike the lead tax, would go to air pollution control research.

Taxation as a tool of government mirrors the condition of government itself, as James O'Connor illustrates. He argues persuasively that the "issue of taxes has always been a class issue." Taxation, whether in the name of "incentives" or of "ability to pay" is today simply a means by which corporate capital escapes the social costs of production by passing them on to the consumer. "Incentives" are justified on the grounds that profits must be maintained for the health of the economy; the "ability to pay" concept on the grounds that everyone receives the benefit of the state more or less equally, and thus everyone should pay in accord with their personal income. The inequities in the tax structure encourage both the exploitation of men and women as well as nature. Because all increased industrial costs are always passed on to the consumer, a rise in the cost of goods will affect first the indigent, and diminish in impact in direct relationship to wealth.

Since 1965 corporate taxes have risen 43.9 percent while individual taxes went up 78.6 percent. In the same period corporation profits increased by over 33 percent while the gross earnings for nonagricultural workers rose by only 15.2 percent, before taxes.

In reality, taxation is always a means of redistributing wealth — putting increasingly more of it in the hands of those who already have most. At the root of industrial responsibility is the profit motivation of production. Taxation in regard to the environment is a means to protect those profits by forcing the additional costs onto the consumer. As such, it leaves the motivation for pollution unscathed.

The continued proliferation of Federal and state agencies on the environment, tax measures, or legislation can only serve to further compound the growing disaster. At worst, support of such activities only lends further credence and authority to institutions incapable of adequate solutions — institutions which inherently foster forms of power and hierarchy which are antithetical to authentic ecological reconstruction. At best, such programs only serve to further obfuscate present conditions.

V

This is equally thc case in other spheres of Federal action. In regard to the urban environment there are a minimum of seventeen Federal departments and agencies, a dozen presidential committees, commissions, and other groups, nearly another dozen coordinating bodies working within the Executive office of the President, countless quasi-official and unofficial agencies and organizations, moving at cross-purposes in some cases, the actions of one body often unknown to the others, seeking to sustain a system of production and power the precise dimensions and costs of which are unclear. In most instances, the banality of the intent is equaled only by the zeal for industrial privilege.

From an ecologically sound political viewpoint, it will not do simply to challenge an occasional project or promote yet additional piecemeal legislation. Quantitative change must give way to qualitative transformation before the whole itself is beyond salvage. What is involved is the very structure of major public service systems (energy, transportation, agriculture, communications), their scale, levels of concentration and consolidation, the kinds of technology developed and suppressed, the virtual impossibility of centralized monitoring or control. Indeed, the very mechanisms of control today work everywhere to minimize the natural diversity and interdependence which are not only the inexorable operating principles of nature but the bases upon which a truly biologically and politically sound society can be built.

III

The Structure of Social and Ecological Responsibility

At the close of the 1960's America was flooded with environmental litter. The magazines, press, bookstores, and media were deluged. From this outpouring we are able to glean a very clear and distinct pattern to the "ecological crisis" so popularly lamented. The most sophisticated formulation of the ideology appeared in *Fortune* magazine's special environmental issue of February 1970. The moral tenor of the issue, paralleling that of the Progressive Era, was set in the editorial which proclaimed, "Unless we demonstrate, quite soon, that we can improve our environmental record, U.S. society will become paralyzed with shame and self-doubt."

Clearly, environmental issues for the mandarins of the economy are questions of morale, a test of the very viability of the machine. An Orwellian-titled article, "How to Think About the Enivironment," asserts unabashedly that environmental issues are simply "unintended by-blows of the modern lust to know and to do." In its many formulations, this description of the "environmental crisis" can be taken as representative of most public statements, whether from industry, government, media, or environmentalists. The destruction of the atmosphere, the contamination of the oceans result from the "unintended by-blows," the inadequate planning, the excessive zeal in the application of technology, inadequate aesthetic sensibilities on the part of architects, planners, engineers, politicians... so the argument runs. In the words of Adam Smith, society "constantly promotes ends which are no part of its intentions."

ii

The question of responsibility is disposed of in almost uniform manner by the "population bombers," Friends of the Earth, the

Sierra Club, and all other conservation organizations. Put in the popular idiom: "We have met the enemy and he is us." The individual is the beginning and the end of the environmental crisis. And yet, nothing could be farther from the truth.

The conception of original sin is rooted deep in the Western mind. The theodicy of evil rests firmly upon the individual. It is as true in present-day formulations by conservationists as in the individualist ethic of James Madison. As John Adams put it early in the American episode, "Whoever would found a state must presume that all men are bad by nature." The implications of this admonition influence virtually everything significant about the form and psychology of our civilization, from the single dwelling unit and form of transportation to the privatization of political participation through voting.

The critical realization that the source of evil in the world resides not in the human gene pool but in the social artifacts of human creation is a relatively recent event. It was perhaps Jean Jacques Rousseau who first demonstrated the social basis of responsibility for evil in the world as opposed to the mythology of personal guilt.[1] The present-day environmental cult of the individual is not unrelated to Enlightenment notions of progress and unlimited faith in the individual who, with good intentions, will "overcome" all obstacles. In recent times, no one personified this conviction more than John F. Kennedy, with his Camelot image. However, as we shall see, the evidence suggests that individuals can no more be held responsible for our present peril than the first dinosaur who walked into a tar pit. It would be to argue that the black person is responsible for the conditions of slavery or the poor for poverty.

The lesson of Adolf Eichmann should impress upon all of us the human vulnerability we all share, and the extent to which the fragile life in which we all partake can at times be corrupted, manipulated, and bent into the most excessive configurations. These configurations result from political and economic choices, choices which, over time, rob the individual of choice itself. This is the significance of the banality of evil.

Institutions often destroy the personality, replacing it with one more tailored, more manufactured to perpetuate the institution.

Today the institutions of everyday life have become the asylums of capitalist society.

One must turn to the institutional fabric behind the forces of production and consumption for the true structure of ecological responsibility. An examination of all basic ecological hazards must seek to identify the true origins of that hazard — how it arose, how it manifests itself today, and what would be required for its curtailment. As we proceed, it should become increasingly clear that the empires of ecological oblivion do not result from individuals and thus cannot be eliminated by individuals. Let us examine a few particulars.

Consumption of electric power is rising; the major impetus for the increased demand comes not from the individual consumer but from rising industrial consumption and increased consumer consumption *promoted by utilities.* Today there are more than two hundred different kinds of electrical appliances available, including refrigerators, hot-water heaters, electric toothbrushes, and knives. The average affluent family has at least twenty such gadgets, many families as many as fifty. Electrical household appliances consume 30 percent of the total U.S. electrical output.

The use of such appliances results in large measure from utility advertising. As *Business Week* stated regarding the energy shortage, "the air-conditioning boom caught the utilities off guard, though they promoted it." One window unit air-conditioner devours the equivalent kilowattage of four-color TV sets. This is more kilowatts than the hot-water heater uses in a household. Contrary to popular conviction, the per capita energy consumption rate in the United States rises five times faster than the population growth. And it is accelerating. If the growth in population were to stop and then recede, gross consumption would still increase.

It is a much quoted fact that everyone in the United States produces six pounds of solid waste per day. This overlooks the reality that many Americans simply cannot afford to consume enough essentials to produce such a quantity of waste. The Citizens Board of Inquiry on Hunger and Malnutrition in the United States estimated that 20 million Americans go hungry every day and that another 10 million people verge on starvation. This is in a country where in 1968 the United States Government paid $4 billion to the agricultural business — the big farmers — to take 35 million acres of soil out of circulation. Since the start of the farm price support system, the large commercial farmers have received $90 billion in price support giveaways. Moreover, only 18 percent of the 30 million poor receive any surplus food or food stamps, and even this is often inadequate to meet dietary needs. Of the families living in poverty, 10 percent are white, 35 percent black, and 58 percent of these blacks are over the age of sixty-five.

In terms of ecological responsibility, nutritional intake is a revealing comparison. While a person living on Riverside Drive in New York will consume a Maine lobster with French wine, his neighbor living in Harlem will eat hominy grits. The simple differential calorie intake should be clear, though often overlooked. The Department of Agriculture admits that *more* people are actually hungry every year in the United States. In 1965 some 20 percent of the diets of Americans were poor as compared to 15 percent in 1955. The vast rise of instant food take-out shops throughout urban America reflects the growing consumption of food with little or no nutritional value. In America every year we consume on the order of 40 billion hamburgers.

Besides the unequal nutritional consumption, there is the fact that the nutritional value of food and other commodities grows geometrically in relation to income. For instance, the cost, capital, and energy of feeding, preparing, packaging, and transporting the Maine lobster as compared to a pork chop is astronomical. The fact that Maine lobsters are packaged and refrigerated to be flown all over the country is symptomatic of the differential energy allocations in the preparation of high-income and low-income food. In Alaska, for instance, almost all the fresh king crab is flown out of that area to the dining tables of more affluent parts of America.

The distribution of capital, or income, is a primary determinant of ecological responsibility. There is a direct proportional relationship between income and the consumption of energy and commodities. This is reflected in the number of appliances owned, the square feet of the dwelling and hence the energy required for heating and lighting, the energy consumed in the care of grounds, et cetera.

The disparities in this land of affluence are severe. The amount of food discarded daily by middle class Americans could feed the total population of many small Third World nations. At the same time, 30 million Americans are daily without enough to eat, not to mention the inadequate nutritional intake of many more millions. At present over 8 million families earn between $3000 and $5000 a year, which is not sufficient to provide minimal daily requirements for food, clothing, housing, and medical care. While 41 percent of the families with an income around $3000 owned a car, all families with an income of $15,000 owned at least one car, 63 percent had two or more. Similar contrasts exist for many other daily conveniences which are taken for granted. Such financial contrasts do not take into account other qualitative differences, such as the additional costs of buying on credit instead of cash, the age of appliances, the quality of basic services, and other disparities between poverty and affluence.

Nearly one out of every four Americans does not enjoy the "affluent life." Absolute wages have declined by 3 percent between 1965 and 1970. The actual percentage of families moving out of the poverty category was in fact greater during the Truman and Eisenhower years than under Kennedy or Johnson. As it turns out, the more wealth one has the less taxes one often pays. The *New York Times,* in an article by Philip M. Stern, reported that 381 Americans with an income of a hundred thousand dollars or over paid no taxes on April 15, 1968. "Twenty-one millionaires pay nothing, while two million persons below the poverty line pay something."

Besides the production of solid waste, the inequities in nutritional intake, and the spiral of added value to high-income commodities, the emission of sewage also reflects the class structure of ecological responsibility. Industrial sewage emissions account for 31 trillion gallons of waste, while all municipal emissions total 14 trillion gallons. These industrial emissions in no way reflect the cost to the water systems involved, since the primary concern is increased profit, not ecological stability.

iii

It should be evident here that the spiral of waste and the production of hazardous chemicals do not result merely from the increased number of persons. While an increasing population contributes to the dilemma, it is a manifestation and not the cause of present environmental conditions.

Put briefly, the increasing number of people in the world today and tomorrow is a unique phenomenon in the evolutionary process. Only man has managed to surpass the life-sustaining capacity of the ecosystem he inhabits. Other animals rapidly achieve a balance between their life-sustaining requirements and the natural capacity of the environment. When that balance is strained, the species suffers and extinction occurs. With man, however, the structure of social organization has been promoted on a global basis which allows him to juggle the natural limits, to import and export natural resources, food, and other vital materials. Though the balance was upset long ago, men and women are just now beginning to realize the cost. In primitive cultures, this balance between men, women, and nature worked unmistakably. The growth of population was harshly limited by the available food. While the growth of population has been geometric in the last several hundred years, one must look beneath that curve for the social forces responsible. For population growth is a reflection not of evolution but social organization. Different forms of social organization obviously yield different rates of population growth. Three stages of human history: the formation of culture, the agricultural revolution, and the industrial age have been the major impetus to population growth.

The gigantic escalation of the world's population corresponds to that period of human history when preindustrial forms of social balance — family and village life, craft, polytechnics, and mutual aid — were destroyed. Correspondingly, sexual relations came to be regarded as a refuge from the toil of industrialization. At the same time, the imperial expansion of Europe systematically destroyed the village character of large areas of Asia and Africa. Without the intervention of Caucasian technology and culture into South America, Africa, and Asia, it is highly unlikely that the current "population explosion" would have occurred. The geometric spiral of birth as opposed to death becomes problematic with the rise of industrial technology and the European explorations.

The imbalance in population growth originates in the imbalance of social organization. As noted earlier, the elimination of ritual and tradition, to be replaced by bourgeois standards of taste, affected every area of life, including the sexual. The social balance of tradition no longer determined the character of sexual activity.

While populations of prior ages steadily expanded, it was only with the industrial age that population began to expand at a rate which prohibited social balance.

What I am suggesting here is that the imbalance between people and resources today is a consequence of earlier social imbalance, the rise of centralized state power. It is precisely the issue of control which is at the root responsible for the present number of people, the control exercised by elite Caucasians upon societies of non-Caucasians. Further control will only aggravate the dilemma. What is required is not control but the reconstruction of balance in the world, between people and resources, among men and women. A society further predicated upon control will only accentuate the imbalances between man and nature. A world order predicated upon "population control" will invariably mean control of rich over poor, white over black and yellow.

While increasing numbers of people may certainly mean increasing levels of consumption for the near future, it is not ordained that way. Given the vast inequities which today exist in regard to the access to resources, balance would dictate some consuming less while others consume vastly more. In other words, a fundamental redistribution of wealth and resources. This means rationing basic, life-sustaining materials to the industrialized part of the globe, and expanding consumption in those areas which now have inadequate nutrition, housing, energy, and the like.

Balance cannot be achieved through further domination. Biological control willl not achieve social balance. The only satisfactory procedure for achieving both social and biological balance is to first promote such balance within a culture, such that the workings of the social order are tailored to the life-sustaining capacity of the ecosystem. No other procedure will suffice. "Population control" such as is practiced by the United States in Southeast Asia, will not achieve world harmony, for it can only serve to inflame already existing inequities.

In the industrialized and nonindustrialized worlds alike, only in a culture which truly values people as people, in which human life is revered and valued as part of the natural web, will attitudes emerge which will allow for the eventual reconstruction of balance between birth and death in the human species.

Only a society in which children are not begat in order to perpetuate the ego of the parents, in which the eternal struggle for the basic necessities of life is abolished, in which the propagation of the species is predicated upon shared and communicated social values, will population balance be achieved. In concrete terms, the objective of those concerned about the growing number of people must be to focus upon the conditions within industrialized nations which promote the imbalance between people and resources. Family planning and contraceptives are no substitute for equality of access to resources or community responsibility for child care. Solutions which continue to require individual participation as individuals will only further perpetuate present imbalances. What the nations of the nonindustrialized world require is political autonomy and adequate resources to promote their health, education, and welfare, not contraceptives alone. Family planning in the technological world, amidst a growing sentiment that the nuclear family is itself a culprit in the increasing propagation of children, will only further the social separation of people which promotes waste and imbalance.

In other words, there are just no technological fixes available to curb the growing disparities between people and resources. What has its origin in a system of political domination can only be eradicated through political liberation. From this the technological apparatus can be evolved to facilitate that which technology can accomplish, not as a surrogate for political liberation.

It no doubt will be a long and tortuous task. In the interim the sheer volume of human suffering will be immense. But that is the price we are now destined to pay for the historic greed of Western civilization. The coming austerity of the United States is a small price compared to the mass suffering in the Third World.

Those who advocate "population control" today are arguing for the further extension of power and control in the hands of those who already exercise it. As such, "population control" in the present historical context will serve to further accentuate those forms of authoritarian control and domination which are at the very heart of the struggle for biological balance on this planet. For these and subsequent reasons, the structure of responsibility for ecological destruction is bound deeply into the social stratification of our society. With wealth comes an increasing surplus value added to the production of commodity items. Wealth, and the social stratification it reflects, is the determining factor in access to resources and the production of waste. Seen in this light, the banal figures about per capita bottles produced, automobiles abandoned, paper thrown out, have a proportionate relationship to wealth.

iv

Within this context we can begin to consider the political arm of the individualistic ethic, the current craze over consumer power. The "myth of consumer sovereignty" has been explored in depth by Ezra Mishan and Paul Baran.[2]

It would of course be absurd to contend that consumers, indeed all who inhabit a society, are in no way influenced by the institutions of that society (schools, churches, politics, culture, advertising) — as absurd as it would be to contend that individuals are completely molded, shaped, and determined by that society. Both such views fail to reflect the interplay which in fact exists between the physiological requirements of each unique person and the existing social environment into which that person is thrust. The issue then reflects the nature of this interaction between the two.

It is extremely difficult to identify physiological traits, to distinguish those which are unique from those which, in most cases, one shares with the other members of the culture and species. We are left then with the task of determining the peculiar mechanisms which capitalist society employs to shape the consumer. How is an individual influenced, shaped, molded, convinced, conditioned?

The conservationists' insistence upon the individual cult of responsibility for ecological imbalances manifests itself in the widespread emphasis upon "what you can do" such as picking up litter, recycling your own litter or newsprint, cleaning up a stream, not driving your auto, or not having children.

In essence this attitude mirrors the capitalist organization of the society itself. Rather than emphasizing the social character of the problem, and thus the social character of the solution, we are told we are to blame, that consumers are sovereign, and that individuals can effect basic change — through the ballot, through a number of individually conscious acts which together, somehow, magically, will change people's attitudes. As we shall see, attitudes and behavior have no such mystical origin.

The environment we manipulate, manipulates us, creates both the possibilities and the limits of organic growth and development. We know that from the moment a child enters this world, the environment of which they are a small but ever increasingly important part, offers possibilities of experience or curtails them. As R. D. Laing has pointed out in *The Politics of Experience,* "Our behavior is a function of our experience. We act according to the way we see things.... If our experience is destroyed, our behavior will be destructive."

While the overt use of force as a means of control is today more apparent, society still relies, in the vast majority of instances, upon the presence of authority for control, or, as Max Weber put it, for the "rationalization of society." Obviously, some patterns of behavior persist, such as sleeping and eating, while most institutions, such as schools and government, are in a state of flux, and others disappear like the Edsel. It is generally the legitimization of authority within an institution, the acceptance related to symbolic status rather than a rationale for actual efficiency or service, which is today responsible for the persistence of most outdated social institutions. While there can be no denying that today legitimate authority is everywhere on the wane, it would be a mistake to understate the hold that the culture still has on most of its children. Today authority is enforced through a multitude of means, ranging from individual conduct and group norms to particular manners and customs of dress. Ritualization is certainly a tool of authority. In order to better understand the means by which a society exercises control and thereby permits or inhibits particular varieties of change, we must understand the patterns of authority present within institutions which permit or deny the question of change or challenge to arise.

Today, while many sectors of illegitimate authority are breaking down, increasingly resulting in the use of force rather than authority to maintain control, the vast bulk of the institutional structure is held together by social norms which are indeed sanctioned by authority of many kinds — church, state, family. More than ever before, these social norms derive their authority from technological systems, integrated into the environment, which solicit passive adjustment. In part, this is because we are increasingly reliant, if not dependent upon, technology.

The extent of our dependence is seldom recognized. Yet almost everything we eat, drink, or use as a tool was produced in a distant place, packaged, and transported, often thousands of miles. Almost everything about our environment, the quality of the air, noise, the temperature of the room, are subject to technological manipulation. Almost without exception, we are completely ignorant of the technological systems upon which our lives completely depend.

As machines are "improved" and become increasingly specialized in function, they permit less and less room for the arbitrary, for what is unique in human beings, in other words, for what is unprogrammed. The movements of a person through the urban maze become increasingly perfected, programmed, steered, until such movements are no longer permitted to express individual personalities. There is no longer room for human error, lest the entire edifice come tumbling down.

The responses of people living in modern societies are mostly reactions to symbolic stimuli. This results in large measure from the expandibility and replication inherent in the forms of mass production that modern technology permits. As the cultural diffusion of technology in a capitalist society increases, technological products become less and less discrete. The similarity of all products in the supermarket or the remarkable resemblance of most inner city cores are examples. The constant renewal and re-creation of commodities and experience provide culturally universal referent images of action and experience. Thus the authority of such behavior resides, without faith or acceptance, in the achievement of something comparable to iconic status through the enormous concurrent repetition and circulation.

This routinization of experience is glibly described by many as elementary adaptive behavior in man. It occurs early with each individual entering our society, and, over time, routinization of experience shapes (or destroys) personality either through the interference with the acquisition of new experience or the determination of certain patterns of responses affecting all behavior. The interplay between a highly regimented physical environment and a highly regimented mentality should be self-evident.

For a society of this order to continue, it depends upon functionally necessary responses. When these reactions are deemed no longer necessary by people — because of biological reactions, social aversion, or for whatever reason — struggle ensues. The advertisement, "Vote! It doesn't matter for whom, just vote!" could easily find its analogy in commodity consumption, "Buy! It doesn't matter what, just buy!" Thus the difference among politicians and between products is minimal. The reaction of students, first at Berkeley, and then almost universally to the IBM cards ("don't fold, bend or mutilate"), typifies the almost instinctual rejection of mechanistic behavior and experience. Herbert Marcuse, among others, has called this the "Great Refusal."

In short, the net result of this conditioning is to foster a sense of adjustment. We learn to adjust. An acceptance-participation ethic is inculcated so that the world is perceived as a mosaic of stimulus response impressions which register as entities in our nervous system and elicit responses. We learn to adjust rather than create. The individual establishes boundaries between himself and the world, a reality he neither helped to shape nor can change. Thus we are taught at an early age that we can do nothing about our environment except perhaps to rebel, for which there are scant rewards. The rewards for the little child refusing to accept the personality-denying authority of the teacher, the student disobeying university regulations, the soldier in the field, or the workers in a factory are scant. This conception of the world as something that exists, to which we must adjust, has profound implications for many aspects of our daily life.

In school, knowledge is presented as something to be acquired, something in existence prior to the learner. History is presented as if a program on a computer tape, as if created by a computer, not subject to error, interpretation, or misjudgment. The young are taught to adjust to historical conditions and thus to the present. History is not something made by people, a process with shifting authentic possibilities, but a predetermined reality.

This ontology carries over into the marketplace as well. We do not question the entire nature of commodity production, but rather individual products. Consumer advocates accept the legitimacy of General Motors' existence, and attempt merely to influence the production of a particular product, or change the composition of the board of directors — almost as though General Motors had attained *a priori* a right to exist as such.

The culturally inculcated separation of the individual from the environment, the fragmentation of the self, the extreme forms of division of labor which exist today under capitalism, are not universal traits of the human species. Rather, they result from a particular context within a particular culture. This was well illustrated earlier by the quote taken from Black Elk. This separation of self, the reliance in the Western world upon the individual as the significant biological and social unit, accounts in large measure for the tremendous emphasis upon "what you can do," emphasizing the individual nature of responsibility and action rather than recognizing both the social character of responsibility and the necessity of social solution. René Dubos has illustrated this lesson, that "man is the only part of the living world which emphasizes fitness in terms of the individual person as the significant biological unit. The most impressive aspect of the law of the jungle is not ruthless competition and destruction, but rather interdependence and coexistence."[3]

The corporate control of production relies in no small measure upon the necessity of individuals — in both their economic and political roles — buying back as individual consumers what the social organization of production denies in the first place. Thus under monopoly capitalism highly sophisticated techniques of market control are developed. The imbalances resulting from capitalist development are attributed to the *conscious* policies of individuals in large corporations, rather than the impersonal forces of the market. Such forces are seen as personal because in that way responsibility can be assigned to individuals "susceptible to change," rather than the impersonal forces of the system *as a whole* — the market. Such an emphasis upon individual responsibility is a method of containment, of directing public attention away from the underlying causes toward the superficial manifestations.

By focusing attention on the individual, capitalism directs public scrutiny toward individual products rather than the purposes of production. Attention turns toward a safety feature on a car or emission standards rather than the inadequate nature of transportation. The atomized public will always pressure for individual commodities. While there may be a chance of obtaining better automobiles, there appears little chance, through individual pressure, of obtaining decent transportation. Collective needs can only be defined collectively.

Andre Gorz recognizes this when he notes that the "preference for the priorities and the values of the 'society of consumption,' for the ideology of mature capitalism is therefore not spontaneous; it arises out of the individual's powerlessness to define and to prefer something else."[4]

In the same sense that regulatory agencies are today a method of maintaining class hierarchy over the means of production, the emphasis upon the consumer serves to insulate the individual from the fact that he is a social producer. Such recognition would be the first step toward liberation. That is why

> the negation of the social origin and character of human needs, the negation of the necessarily social mode of their satisfaction, and the affirmation of the possibility of individual liberation through the acquisition of the means of escape (whose social production is carefully masked), are the fundamental mystifications of the so-called affluent civilization.[5]

The designation "consumer" immediately reveals the source of the problem. We are taught that individuals can influence the political economy first and foremost as consumers, that is, as economic entities in regard to the cash nexus. Participation in and contribution to the commonweal is primarily in the marketplace, primarily through the relationship to the means of production and consumption. Individuals at the outset are cast into a single category as actors in the drama of commerce.

Commodities thus assume the life that human intercourse denies. This is what Marx understood by the "fetishism of commodities" in the first chapter of *Das Kapital.* The fragmentation and alienation resulting from the division of labor leads to a life of fantasy which resides in commodities. The interplay between production and consumption is critical for understanding the ecology of capitalism. Again, citing Marx: "Production furnishes consumption with its material object.... It gives consumption its definite outline, it character, its finish.... Production not only supplies the need with an object, but also supplies the object with a need." In other words, production both serves those needs dictated by the market, and also creates markets which in turn generate production.

Nothing better demonstrates this than advertising. We in the United States are faced with a mode of production and consumption, bound one to the other, which squanders over $20 billion a year to promote products designed primarily for marketability rather than usefulness. They are produced not so much for the gratification of the consumer but for the producer — for the increasing profit margin of the industry. The scale of this sham is staggering. In 1971 the Food and Drug Administration posted a list of 369 drug products which were either ineffective or hazardous. This represents 12 percent of all the new drug items introduced betwen 1938 and 1962!

One example of this sham is Alka-Seltzer, a household standard. Miles Laboratories, the producers, spend $23 million a year on Alka-Seltzer advertising alone against sales of $60 million. Their total advertising budget of $45 million is five times the research budget of $9.6 million. Listerine, had sales in 1969 of $70 million with advertising costs of nearly $20 million. How much does it cost to produce mouthwash?

Often drugs and cosmetics run as much as 30 percent or more of their costs in advertising. Chevrolet, the most advertised item in the economy, spends $90 million a year or $45 for every car sold.

These are illustrations of the added value which most consumers pay for in the market. Style changes, built-in obsolescence, advertising, credit, built-in faults, excess packaging and the like cost billions every year. Such a form of production sacrifices the welfare of the many for the profits of a few. The corporate organization of profit directs the society, instead of society democratically directing the means and ends of corporate organization. We must struggle toward a society in which human beings as such are sovereign — not consumers, producers, or managers.

v

The conditions for such struggle will require a fundamental reconsideration of our present notions of the nation-state and the role and influence of America as a world power. The political limitations of nations do not apply to the natural interplay of environmental systems. While governments have one set of demarcations for the globe, nature has quite another. America's global ecological influence results both from activity undertaken within the continental United States as well as foreign activity. Both must be considered the ecological policy of America.

A decade ago scientists in the Antarctic were amazed to find residues of DDT in the fatty tisues of birds who lived thousands of miles from any farm. Similarly, in October 1970, scientists found that the livers of Alaskan fur seals that live in the open Pacific scores of miles from land, contained amounts of mercury at levels considered toxic to humans. The Food and Drug Administration ordered the removal of 10,000 iron supplement pills made from seal liver. Concentrations of industrial and agricultural chemicals, in the case of these birds and seals, were passed through the food chains. We can conclude from these examples and scores of others that virtually no spot on the face of the globe escapes the biological hazards resulting from advanced industrialization. The United States, without comparison, leads the world in the production of agents hazardous to the biosphere. The United States, for example, produces two-thirds of the world's pesticides within the United States, over one-half of which are exported. This does not consider that fraction of the remaining one-third produced in other countries by U.S. companies. Today America exercises a vast influence on the course of social, technological, ideological, and ecological events on a planetary scale. It is impossible to neatly separate these sectors of a society, for we find that the activities of America in the course of influencing the military, political, and economic priorities at home and abroad all have profound ecological implications on a scale beyond normal comprehension.

We persist in viewing America through nineteenth century glasses. The confines and limitations of a nation-state have long ago been surpassed by the United States. America is not a nation at all, but an empire. The natural borders of water and land exercise virtually no influence on the global impact of America — whether in terms of the obliteration of subcontinents like Indochina, the universal spread of atomic fallout, the worldwide spread of mercury, or the vast social influence of trade and ideology. The ecological imbalance which has by now been well documented results directly from the social mode of organization which is the origin of waste, hazardous chemicals, and the interruption of ecological systems. There can be no way to understand the global disruption of the biosphere without understanding the global organization of political and economic activity. We must begin with an understanding of the nature of production itself.

The capitalist organization of production requires the increasing involvement of the state in stabilizing the means of distribution. From the time of Hamilton the State has become increasingly active in tariff and trade regulations. We have considered earlier the ways in which regulatory agencies of the state are motivated by, and become servants of the class organization of society. The increasing arena of foreign investment is directly tied to the class orientation of capital.

> ... the attempt of regulatory agencies to maintain profitable conditions in a particular industry tends to freeze the pattern of resource allocation, establish monopoly conditions and so on, which in turn retards capital accumulation and expansion in the economy as a whole. Foreign economic expansion thus becomes increasingly important, as a key mode of economic

> growth and as a way to transform interest group conflict into interest group harmony. And foreign expansion clearly requires a class conscious political directorate.[6]

We can thus understand the increasing preeminence of America in the drive for international market control. The resulting tendency toward the elimination of competition on an international basis makes for the more efficient distribution of natural resources, capital, labor, and new markets for commodity items.

The Ford Motor Company, for instance, exemplifies the aggressive international corporation. The sixth Ford car built was shipped to a Canadian distributor. Ford in its first year of operation started making arrangements for building up its foreign markets. The much noted visit in 1970 of Henry Ford II to the Soviet Union dramatizes the situation, even though Russia rejected Mr. Ford's proposal. William O. Burke, president of the newly formed Ford Asia-Pacific-South Africa, Inc., suggests, "If only we can get just one percent of that market it is mind blowing.... Frankly, I look forward to the day when there will be a Ford Motor Company in China." He was, no doubt, reflecting the aggressive interest of many companies to trade with Communist China — though it seems hardly likely that the Chinese would be interested in spreading the internal combustion engine around China, particularly American made. While the imperial adventures of American corporations are certainly not a recent development, it is only since World War II, that the United States has exercised such a vast preeminence. The acceleration of foreign investment is staggering. By 1961, some 460 of the largest 1,000 U.S. companies had a subsidiary or branches in Europe; four years later in 1965, there were 700 out of 1,000 largest U.S. companies in Europe.

At present, there is an estimated $400 billion worth of goods and services produced through international investment — half of which is American. Half of all the goods and services produced in the world are American. Many American corporations derive a large part of their profit from foreign investment. Standard Oil of New Jersey, for instance, earned 54 percent of its 1969 profit from foreign investment. Such multinational giants — particularly those involved in ecologically popular industries (such as petroleum and automobile production) — come to rely increasingly less upon domestic production and sales, and are increasingly less susceptible to "consumer pressure."

Foreign investment exhibits an ever-increasing concentration of capital resources. In the three largest European markets, West Germany, Britain, and France, 40 percent of the U.S. direct investment is accounted for by Esso, General Motors, and Ford. Considering all of Western Europe, only twenty American firms represent two-thirds of all U.S. investment.

This international situation is a reflection of the conditions at home. We noted earlier that *Business Week* estimates that the 200 largest industrial corporations control over two-thirds of the nation's manufacturing assets. This means, for instance, that General Motors, AT&T, and Jersey Standard account for 6.2 percent of total domestic U.S. sales, that is a 20 percent boost over the last ten years. Economic concentration is increasing. Attorney General Mitchell estimates that were it not for recent mergers there would have been 600 more companies in the last decade with assets of more than $25 million, or nearly 50 percent more than the present. Mergers reached their height in 1968 when thirty manufacturers with more than $100 million each disappeared from the market. The severe economic concentration within the energy industry will be considered later.

This process of consolidation depends upon growth as well. The acceleration of the economy over recent years far outstrips the pace of population. It took until 1830 for the economy to reach a GNP of $1 billion. By 1880 it reached $10 billion, 1929 upwards of $100 billion, sloped down in the depression until it recovered at $100 billion again in 1941. By 1949 it had jumped to $250 billion, $500 billion in 1960, and now surpasses one trillion dollars — it has doubled in ten years. The GNP of the United States is three times that of West Germany, France, Italy, the Netherlands, and Belgium combined. It is five times that of Japan. The implications for ecological responsibility are profound. Just since 1955, Americans have doubled the number of cars, tripled the electric power consumption, and increased tenfold the consumption of plastics. In terms of average per capita toxic waste that flows into rivers and streams, each American produces one thousand times more than an Asian.

The sheer size of the American corporation as an economic unit has global implications for resource consumption, the transportation of goods and services, overseas offices, the balance of payments of nations, trade agreements, et cetera.

Dominant American corporations compare rather favorably with many nations in terms of GNP. Of the largest 50 economic entities (in the world) about 37 are countries and 13 are corporations. Of the top 100, 51 are corporations. Such corporations, most all of which are to be found in the top 100 *Fortune* directory, have double the growth rates of many nations. One acquires some sense of the scale involved by realizing that California, if considered as a nation, would rank seventh in terms of GNP. California has as many autos as the Soviet Union and has more telephones than any nation except the U.S. and Japan.

The massive acceleration in the accumulation of capital, or GNP, differentiates the $100 billion economy of 1941 from the $1 trillion economy of today. Prices in this period have soared, doubling between 1941 and 1961. From 1961 through 1969, prices have increased by 30 percent. The expansion of income and market entails many ecological implications. Mass production and scale require large inputs of capital. All forms of transportation are increased to accommodate the flow of goods and materials. Under capitalism increasing production inherently entails increasing magnitudes of waste and surplus. Increasing efficiency results in ever higher levels of profit. Real GNP has increased tenfold since 1900. About half the increase in output can be attributed to economic efficiency, given a somewhat stable supply of land and labor.

"Efficiency" is the motor force of the American economy. In theory, the market is the test of efficiency. Resources are allocated efficiently when they maximize one's position in the market, in other words, when they increase profit. Increasing profit of course functions to stabilize the market, provide for predictability, and reduce competition and waste. Efficiency can thus be equated with control, with predictability, et cetera. In order to allocate investments and resources in a rational manner, corporations must achieve a somewhat predictable market situation. Control over the market is the overriding thrust of efficient management. Efficiency and control may be considered the same under capitalism. That is because the overriding principle of social and economic organization is to further enhance the division between the social organization of production and the private organization of profit.

vi

There are innumerable ways in which the accumulation of capital and the drive for efficiency and control have a profound influence on the ecological stability of the planet. Capital is the critical factor in determining development, urbanization, methods and amounts of production — in short, the very structure of industrialized society. As we shall see, without vast amounts of capital and control one cannot assemble the technological mechanisms required for ecological imbalance on a truly proper global basis. The contrasts in GNP are suggestive of political contrasts regarding ecological responsibility. One example of this process is urbanization. Of the twenty-five largest cities in the world, six are in the United States, and only one other country has more than one. Yet the United States comprises only 6 percent of the world's population. No other nation occupies such a preeminent position in the encouragement of urbanized congestion as does the United States. As a consequence of American influence the multimillion-person city is the fastest growing form of human settlement. In 1920 some 35 million persons lived in cities, in 1960 over 169 million, a five fold increase. It is, by and large, American economic activity, development, and planning which are responsible for the rapid urbanization in many parts of the capitalist world. The influence of such American firms as Skidmore, Owens, and Merrill, or of someone like Buckminster Fuller in influencing the global course of urbanization is critical. They merely manifest the impact of American international investment.

The magnitude of economic investment abroad brings with it innumerable implications for the development of the country involved. Concentrations of capital (in many cases by corporations or their subsidiaries which rival the economic output of the nation they are investing in) bring with it the concentration of manpower, machines, technology, transportation, and communications. One need only visit any major city in the world to observe the dominance of American advertising and media in regard to commodity consumption. Another indicator of the rapidity with which the Pax Americana has penetrated the international capital market is the extension of U.S. bank branches. By 1960 there were branches in 33 nations, by 1967, at least 55 nations. In terms of the number of branches involved, there were 95 in 1950, 124 in 1960, and by 1967 there were 298 branches of U.S. banks abroad.

With American involvement abroad comes the attendant "American style of life." One particular agency of American-induced urbanization is the international hotel chain, foremost among which is Hilton. American "inroads of glass and steel" now mark the skyline of virtually every major city in the world. The top ten American chains boast 369 major high-rise hotels either in service or in construction as of 1969 with Hilton leading the line with 63. The "jet age hotels" are fed by an international jet set class — pushing American patterns of consumption and waste into every society of the world. This culture is, of course, supplied by American-produced jets, from Boeing. At any time of day, there are nearly one quarter of a million people in the air on American-made jets — contaminating the air, congesting the transportation corridors on the ground, generating noise — in short, all the attendant ecological hazards of airlines. The introduction of the 747, for instance, forces the airports and air industries of the world to virtually adapt to scale their airlines and airports to meet the dimensions of the American-introduced air travel.

With American culture come American commodities. Foremost — aside from the ever-present Coca Cola signs (Coke with the capacity of taking the paint off your car, let alone the effects to your stomach) — is the automobile. One illustration of American responsibility for the global destruction of the atmosphere is the number of automobiles in use around the world. The *United Nations Yearbook* for 1969 records that in 1968 the United States produced 38 percent of the world's automobiles and trucks and employed 45 percent of all motor vehicles. Considering American automotive production abroad, wholly or partly owned, we might estimate that American capital is responsible for about 70 percent of the world's motor vehicles.

We will consider the political economy of automobiles and petroleum in a later chapter but here we may just point out the staggering global ecological culpability of the American automobile. Consider carbon monoxide (CO). Carbon monoxide is the principal contaminant. It results from the incomplete combustion of carbon in the presence of oxygen which occurs primarily in the burning of fossil fuels inside the internal combustion engine. Lesser amounts of CO are produced in the combustion of coal, in furnaces, some industrial operations, the disposal of solid wastes, and natural fires. However, over 60 percent of CO emissions from all sources in the United States alone result from the internal combustion engine. Total global emissions are estimated at 257,000,000 tons.

Carbon monoxide is considered toxic to humans at concentrations of 100 parts per million with exposure for several hours, and lethal at concentrations of about 1,000 ppm. Recently the allowed industrial exposure to CO for eight hours was lowered from 100 ppm to 50 ppm. This is on the low side of the average exposure when driving on the freeway of any major city in the country. In Detroit, during a short peak, the ppm count has been recorded as 100 ppm. Some of the symptoms of 50 ppm include a worsening ability to detect a flashing light against a dim background and some impairment of ability to judge time. While there is no conclusive evidence to link such CO concentrations with impairment of driving, it is certainly possible. It should be evident that the internal combustion engine, produced and distributed almost exclusively by the United States, is the single most important source of CO. It is estimated that 95 percent of all CO is produced in the northern hemisphere. There can be no doubt that most internal combustion engines in the hemisphere are by and large American produced. Anyone ever visiting Mexico City cannot help but observe and smell the extremely high CO emissions resulting from the large number of very old, American-made automobiles.

Besides the emission of CO, the internal combustion engine is responsible for nearly 60 percent of other emissions into the atmosphere including hydrocarbons, nitrogen oxides, sulfur oxides, and particulates. In a number of places these emissions from high technology sources have resulted in thousands of *recorded* deaths, and untold others. Recorded instances include Belgium's Meuse Valley (1930), Donora, Pennsylvania (1949), London (1952, 1959, 1962), New York City (1953, 1962, 1966, 1970). However, in the 1953 incident, the consequences were not really recognized until nine years later when someone did a comparative analysis of hospital records.

By and large, the medical profession and the public health services are totally unprepared to examine the relationship between air contamination and premature death. This would involve a fundamental reconsideration of the origins of death and disease in the urban complex, the extent to which death is socially determined, and to which politicians rather than physicians should be responsible for innumerable forms of urban pathology resulting in death and disease.

Air contamination is now directly linked to an increased incidence of cardiovascular disease, strokes, cancer (particularly bronchogenic cancer), bronchitis, and emphysema. The number of cases of emphysema in the United States has doubled every five years for the past two decades. The automobile is but illustrative of the pattern of responsibility regarding the contamination of the atmosphere. Industrial chemicals and radiation have similar effects. In short, the chemical balance of the gases in the earth's atmosphere is today largely a consequence of American-owned or induced production.

Carbon monoxide is but one gas that is largely the production of American enterprise, carbon dioxide (CO_2) is another. The estimates regarding the human-induced increase in CO_2 in the atmosphere since the beginning of the century vary from 10 to 14 percent. While CO_2 occurs naturally — largely as a result of animal respiration — the largest source in the atmosphere today is from the internal combustion engine and the production of electric power. We have already noted the U.S.'s responsibility for the automobile, but the U.S. also consumes and produces over one-third of the world's total electrical output. There is abundant speculation that the rise in CO_2 will affect the earth's temperature over time. As Francis S. Johnson of the Southwest Center for Advanced Studies in Dallas remarked at a symposium by the American Association for the Advancement of Science, "The risk of a serious perturbation appears to be small, but the problem is poorly understood and the confidence level in such a prediction is low."[7] At the same conference, an official of the U.S. Department of the Interior noted:

> It is obvious that man can shift the natural equilibrium forces that make for a healthy balance in nature essential for life, in particular the composition of the atmosphere and the food potential of the land and the sea. Thus far a major disruption in the atmosphere is not apparent. We may, however, be looking at subtle changes in the environment, such as the accumulation of lead and the other toxic chemicals, and increase in carbon dioxide and carbon compounds, atmospheric dust, and the acid rainfall which may have a far reaching consequence in the years to come.[8]

The above two citations are from a book edited by Fred S. Singer, *Global Effects of Environmental Pollution,* a symposium conducted in 1968 by the American Association for the Advancement of Science. To my knowledge, this is the most reliable and most comprehensive source of current knowledge about the effects of American society upon the biosphere.

One estimate suggests that in comparison with the rest of the world, the United States is responsible for one-third to one-half[9] of the production of hazardous waste products. The following chart estimates U.S. domestic emissions as compared to global emissions. As a consequence, the U.S. figure, if one accounts for U.S. emissions produced abroad, would be considerably higher, considering that the United States produces one-half the total goods and services produced abroad. The extremely precarious state of knowledge regarding the effects of such emissions is summarized by Francis S. Johnson:

> Because of the importance of these problems to man's future, we should be very confident of all our understandings of them. At present, the ideas and even the numbers must be considered highly conjectural. . . . One of the many possible ways in which pollution can precipitate a catastrophe is by upsetting the oxygen balance in our atmosphere. However,

Estimated rates of injection of materials into the atmosphere (gallons per year):

Material	*U.S. Rate*	*World Rate*
CO	7×10^{13}	2×10^{14}
Sulphur oxide	2×10^{13}	8×10^{13}
Hydrocarbons	2×10^{13}	8×10^{13}
Nitrogen oxide	8×10^{12}	5×10^{13}
Smoke partic.	1×10^{13}	2×10^{13}

SOURCE: Fred S. Singer, *Global Effects of Environmental Pollution* (New York: Springer-Verlag, 1970), p. 179, from a paper by Edward D. Goldberg, "The Chemical Invasion of the Oceans by Man."

the time constant for this is sufficiently long that a destructive course may well have been followed beyond the point of no return *before it is recognized* [emphasis added].[10]

Let us carry the question of responsibility one step further by considering the chemical invasion of the oceans by man. Perhaps the first indications of what is transpiring just below the surface of the world's oceans is what is happening above — the disappearance of the bird. In California, the pelican, long a symbol of the North American coast, acquired so high a concentration of DDT in the shell of its egg that in 1969 not enough pelicans were hatched to perpetuate the species.

A reduction in carnivores, particularly carnivorous birds, is an early sign that the structure of natural communities is upset. Because of their high metabolic rate, birds consume quite a lot in proportion to their body weight. Consequently, drastic changes in populations are readily visible.

With an annual release of 200 million pounds of DDT into the biosphere, an equilibrium concentration is estimated only after seventy years, given an estimated ten-year half-life for DDT residues. George M. Woodwell concludes that "there is reason to expect, then, that DDT alone will account for a significant degradation of oceanic system, including the oceanic fisheries, in the next decades unless its use in places where it can contaminate the living systems of the earth is halted."[11]

The chicanery of the Department of Agriculture seems to preclude any significant curtailment of DDT. Today each person in the United States averages 12 ppm of DDT in their body— an amount the FDA considers unsafe to cross state lines in products.

We should note at this point that the United States accounts for 23 percent of the production of nitrogen fertilizers in the world, twice the production of the USSR and three times that of Japan. While the sale of DDT has been officially banned for use in the United States, Thomas Whiteside reports in the June 18, 1970, issue of the *New Yorker* that it continues to be sold almost without exception. Here again the relationship between war and industrial production: DDT was first developed for the military in World War II.

DDT is, of course, merely one chlorinated hydrocarbon among an entire variety of substances which are daily disposed of in the ocean's sink. The transfer of mercury from the continents to the oceans via the rivers is estimated at five thousand tons per year. This is about half the annual total production of mercury. The vast bulk of the lead entering the atmosphere and oceans results from the associated uses of the automobile. Antiknock lead was first produced by General Motors.

Man appears to be responsible for an input [of lead] into the oceans equal to that of natural processes. In the Northern Hemisphere about 350,000 metric tons of lead, as the antiknock agent lead tetraethyl, are burned in automobile and internal combustion engines and subsequently introduced into the atmosphere... about 250,000 metric tons of lead are annually washed out over the oceans and about 100,000 metric tons over the continents in correspondence to their relative areas.

> This impingement by man has raised the average lead content in the surface waters of the Northern Hemisphere oceans from about 0.01-0.02 to 0.07 micrograms/kg of sea water in the 45 years since the introduction of lead as an antiknock chemical.[12]

A report prepared in 1970 by the Dillingham Environmental Company of La Jolla, California, "An Appraisal of Oceanic Dumping of Barge-Delivered Liquid and Solid Waste from U.S. Coastal Cities," identified 126 U.S. ocean-disposal cities — 42 in the Pacific, 51 in the Atlantic, and 33 in the Gulf of Mexico — in 1968. Only 20 coastal cities used the ocean for dumping in 1968. This is by no means a complete indication of present U.S. dumping procedures. Many industries employ private disposal companies for ocean dumping, avoiding closer monitoring by public officials. This report indicated that the factor most responsible for the increase of tonnage over two five-year periods was industrial wastes.

Aside from the regular industrial wastes that are often highly toxic, the U.S. supplies the oceans of the world regularly with atomic wastes, chemical and biological warfare agents, explosives, and a large amount of crude petroleum. Annually somewhere on the order of 10^{13} gallons of gasoline and 10^{12} gallons of various solvents are dumped into the oceans. The instances of U.S. life-destroying production are legion. The daily press monitors the flood of toxins into the biosphere. But perhaps the key to understanding the ecological implications of a trillion dollar economy is the production and consumption of energy. Energy is what makes possible the magnitude of mass production and the astounding economic growth. Indeed, perhaps the most critical confine of the market economy is not capital but available energy. The results of energy failures in the Northeast dramatically demonstrate this point. Without massive energy production, the substitution of capital for labor would not be possible in industries such as petroleum and chemicals.

In 1968 the total world consumption of energy in millions of metric tons of coal was 6013.67. Americans consumed over one third, or 2078.17.

Consumption

Total World: 6013.67	U.S.	2078.17
	Europe	1159.81
	Asia	435.36 (half of which was Japanese)

Kilograms per capita

World Average: 1733	U.S.	10,331.00
	Europe	3,309.00
	Asia	406.00
	Japan	2,515.00

SOURCE: *UN Statistical Yearbook,* 1969.

These figures do not include the consumption of energy by U.S. industry abroad, or the military. The ration of energy consumption is the most important indicator of ecological responsibility — for energy is the key to the production of hazardous chemicals, gases, and solids which are daily poured into the life support systems. There is a direct correlation between a rising economic development and a rising consumption of energy. However, the difference between the contribution of an American and his Asian counterpart is not simply one of magnitude — but a qualitative differential. Asian society is not organized per se, around the production of surplus and waste; it is a society of scarcity. The consumption of energy reflects an organization of scale which qualitatively separates the American mode of production and consumption from that of every other mode in the world. That is not to suggest that the Soviet Union, Japan, West Germany, and other nations do not produce surplus, waste, or pollution. The difference between the United States and these other nations is qualitative, not quantitative, similar to the difference between the atomic bomb and conventional explosives.

There are numerous implications of the energy ratio — such as the inequities in depletion of natural resources — in terms both of fossil fuels and minerals utilized for manufactured goods and services. Water, for example, is required in vast amounts not only for agriculture, but all industrial processes. One gallon of petroleum requires 7-25 gallons of water for processing; one ton of steel about 25,000 gallons; a ton of copper, 50,000 gallons; a ton of acetate, 250,000 gallons; and a ton of synthetic rubber requires almost 600,000 gallons of water.

Inequality in food consumption is perhaps the most popular example of resource depletion and waste. George Borgstrom in *Too Many* put it this way: "Basically there are not many oases left, almost worldwide network of slums, about 450 million well-fed people living in comparative luxury... as against 24,000 million undernourished or in other ways deficiently fed and generally poor."

There are by now several works which document the imperial adventures of the United States in search of raw resources abroad. Brief examples should illustrate the situation. The United States in 1966 accounted for well over one-third of the world's consumption of tin, one-fourth of its phosphate, potash, and nitrogen fertilizer consumption, half of its newsprint and synthetic rubber, more than one-fourth of its steel, about one-fifth of its cotton, one-third of the electricity, and as we noted earlier, practically all the automobiles, and one-third of all the paved roads in the world.

The present glut of American consumption must be severely curtailed. If everyone in the world were to demand present American standards, seventy-five times as much iron, one hundred times as much copper, two hundred times as much lead, seventy-five times as much zinc and two hundred and fifty times as much tin annually would be required.

vii

These observations are designed to illuminate the structure of ecological responsibility in the world today. The mechanisms which determine the production of dangerous chemicals and other hazardous substances are not found in "consumer power" or simply the greedy nature of the American people. It should be clear that the vast proportion of atmospheric contaminants (95 percent of the global CO for instance) are generated within a very small fraction of the globe, those areas under direct control and manipulation of America. We must realize that both military and economic activity of America abroad is motivated toward a single and all-encompassing objective: control.

Individuals exercise relatively little if any influence on the course of production and consumption any more than they do upon the course of political decisions. The failure to recognize the structure of control and domination in both the physical and social dimensions is today a commentary as much about Ralph Nader as it is about J. William Fulbright.

The imbalances generated between social organization and natural evolution have achieved their ultimate expression in the United States. This contradiction is well illuminated when one realizes that the consummate increase in "real wealth" (GNP) is today everywhere accompanied by a decrease in social welfare. While capital accumulation increases with vigor, life becomes increasingly less satisfactory. Not merely to the factory worker or the ghetto resident, but to members of the white middle class who must commute hours daily, or for the corporate executive whose social service systems are failing.

The interrelationship between social and biological imbalance cannot be overstated. The ecological effects of present industrial organization manifest themselves as much in the health of the population as in the balance of life-giving processes. Blue Cross in a 1969 publication entitled, *Sources,* indicated that more than one out of every ten families in the United States has a seriously ill member. They further stated that 60 percent of the total public feel worried or nervous, 52 percent sometime feel lonely and depressed, and more than a quarter of the population are sometimes unable to sleep, to stop smoking, to eat, and to get up readily in the morning. In isolation such statistics are meaningless, but taken together they reveal a constellation of pathology which affects nearly everyone.

The drastic imbalance between poverty and wealth, the consumption of liquor and cigarettes compared to the expenditures on health care (the average family allocates $125 annually while the average hospital stay cost $550 in 1969), are recognized as classic symptoms of capitalist development. Robert Heilbroner, in a paraphrase of Ernst Mandel, has captured the essence of the issue:

> The private form of appropriation makes profit the only aim and driving force of production. It causes the development of productive forces to be uneven. Production develops by leaps and bounds, not in the sectors where the most urgent needs are to be found, but rather in those where the highest profits are to be achieved. The production of alcoholic drinks, of "comic books," and of drugs takes precedence over the struggle against air pollution, the preservation of natural resources, and even the building of schools and hospitals. Underproduction in one branch regularly coincides with overproduction in another.[13]

The merger of state and corporate capital is the mechanism to facilitate this gigantic productive sham. In earlier times the state did not provide massive fiscal services to corporate enterprise, such as regulation, testing, research, and development funds. The development of productive forces and the concentration of capital represent a new stage in the development of capitalism: monopoly capitalism. The combination of corporate and state capital, fully integrated, affects life both biologically and socially.

This integration of capital has proceeded apace in virtually every sector of the society. We have seen how the momentum for Federal regulation by the conservation movement during the Progressive Era did much to facilitate the consolidation and integration of the public and private sectors. Today this integration pertains not only to the production of commodity items but to the production of knowledge and ideas, culture, and personality itself. It becomes increasingly incumbent upon political capitalism to manage reality on a scale unprecedented during the nineteenth century. In such an atmosphere, the state becomes the vehicle by which to socialize the costs of the private organization of production.

Today these costs, or what some economists call "externalities" are everywhere to be observed. "Externalities" in terms of the maximum excess value are added to every commodity. Lead is added to gasoline in order to increase octane, or power in the vehicle. DDT residue is utilized in order to increase the "efficiency" of agricultural production which in reality begins to destroy the consumer himself. Oxygen is utilized in transportation . . . et cetera. In short, industry is not required to account for the resources they despoil in the process of production. No responsibility is taken for the maintenance of the natural order because there is no capital provided to do so. This is what Marx meant when he said, "Private property has made us so stupid and partial that an object is only ours when we have it, when it exists for us as capital or when it is directly eaten, drunk, worn, inhabited, etc., in short, utilized in some way."

Put still another way: "all natural relationships have been dissolved into money relationships." Today the ownership of property and capital is equivalent to the ownership of the environment itself. The concentration of production and capital, of industrial capital with banking capital, the increasing rise of exports over imports, the increasing production of surplus in relation to total product (waste) — all combine in the ultimate concentration of the multinational corporation. In its consummate development, capitalism requires constantly accelerating levels of consumption, rising productivity, and the impulse to accumulate for its own sake. Marx called it the circulation of capital. We call it growth. Growth through the multinational corporations involves multiproduct, multiindustry, multimarket, and multinational organization. As Dr. Edwin G. Nourse, first chairman of the President's Council on Economic Advisors testified before a congressional committee in 1963: "There are no demonstrable or discernible limits at which such a concentration of economic power, once fully under way, would automatically cease." Our observations earlier about the sheer size of such multinationals demonstrate this admirably. In such an atmosphere, we are not talking about the divvying up of the world pie, but about the process of distribution; the pie is owned, operated, and controlled by American capital. In 1965 the *Economist* of London estimated that of all the giant corporations with more than $700 million, the U.S. had 272, Britain 54, Japan 30, and the rest of the world 99. This magnitude of capital organization reflects a multitude of commodities, each as uniform as the order which produced them. Waste, for its own sake, or rather for profit's sake, becomes the measure of production. The development of surplus, or waste, was a primary theme of *Monopoly Capital* by Paul Baran and Paul Sweezy. They estimated that in 1963 the surplus as a percentage of the total GNP amounted to 56.1 percent.[14]

This organization of production for commodities which are most profitable rather than most needed by people is not accidental. It is what Andre Gorz has called in a *Strategy for Labor,* "a model of affluence" which levels consumption "upward." In other words, the primary objective in the development and marketing of a product is the maximization of "added value" through excess packaging, style alternation, sexual association, class association, and a multitude of other psychological mechanisms employed by major marketing departments. "The tendency to prefer the accessory to the essential, the improvement of the profit rate to the improvement of use value, has resulted in *absolute* wastage." The production of waste is thus a technique employed to obtain ever increasing amounts of capital by producing objects — without actual use value — designed to soak up the excess capital of the affluent consumer. This mechanism functions as much to the detriment of the environment — objects discarded, maximum turnover, non-recyclable packaging — as to the detriment of the social order — few get richer and most get poorer.

But perhaps the most pressing ecological concern here relates back to our earlier discussion about the myth of consumer sovereignty. Namely, that while the environment is deprived by people through the *social* forms of organization, we are expected to recapture the oxygen in the atmosphere or the clarity in the water on a *personal* basis. Or, again in Gorz's terms, ". . . The nature of capitalist society is to constrain the individuals to buy back individually, as a consumer, the means of satisfaction of which the society has socially deprived him."[15]

This profound fragmentation of society has its economic as well as political advantages. The maximization of added value depends upon the organization of the society in terms of individuals, for example, the use of automobiles instead of collective forms of transportation. In the social dimension, the division of labor functions as a mechanism of control, both over the means of production itself and over the individual worker, in terms of taxation, wages, and the other conditions of everyday life. The regimen of the military is adopted in the factory. Men and women become as replaceable as the commodities they are producing. Such standardization makes for increasing efficiency, and thus, increasing profit. In these and other dimensions we must come to understand the nature of the productive apparatus of America as fundamentally tied to the organization of the society as it now exists; to the increasing isolation of individuals, both as producers and consumers, to the increasing turnover of commodities, styles, and fashion, and to the maximization of waste or surplus which has a direct bearing upon the stability of the life support system.

This surplus functions first and foremost to reduce diversity. These mechanisms today have achieved a destructive capacity unknown before to mankind. In short, the life support system of the earth can no longer sustain the forces which the expansion of the modes of production and consumption have assumed. The trillion-dollar economy brings with it a structure of commodities which requires the fantastic production of dangerous chemicals, surplus packaging, solid waste, and effluents which are incompatible with the life-sustaining capacities of the planet itself.

These issues will not be addressed by some public recycling campaign for Coke bottles or newspapers. For we are dealing here with an infinite expansion of consumer durables, each of which reflects the very structure of production. To deal with any single item alone is merely to bolster the system itself. We observed earlier how Federal regulation, or for that matter regulation on any governmental level, lends credence and legitimacy to the commodity or service itself, never calling into question the very viability of the production apparatus.

These inbalances are paid for in the economic sphere primarily by you and me, as individuals, rather than by the social organization of production which is responsible. What is involved here is not reducing the profit margin of commodity production — such a rational step is not even considered by corporations — but rather passing the "costs on to the consumer." We must address not simply the proportion of profit but the system which assigns profit priority above community control. The present schemes of consumer power and social taxation attached to commodities to pay for "pollution control" merely serve to bolster the system of commodity production which is at the root of the present dilemma.

The rationality of production functions as an integrated system of domination. This system operates universally so that the domination of man is inextricably bound to the domination of nature in all aspects of our everyday life. Today the logic of social and ecological imbalance is rooted, more than anywhere else, in war and the petroleum and automotive industries. Bertrand Russell recognized this in *Liberty Against Organization* when he said,

> Two men have been the principal creators of the contemporary world: Rockefeller and Bismarck, the first in the economic world, and the second in the political, [they] destroyed the liberal dream of happiness of all inhabitants of the planets . . . substituting the monopolistic organization and the corporate state.

IV

The War Machine

From the sixteenth century through today, war has provided the basic model and impetus for mass production and mass consumption. The standardization of production required to sustain the army became the standard mode of production in "peace." The first mass produced commodity was the uniform for the armies of Napoleon. Long before the industrial revolution, mass production was established in the arsenals of Venice.

For four hundred years the regimen of the military has dominated the fabric and purposes of industrial society. All aspects of civilian life — industrial invention, the standardization of production, appearance, behavior, and time — are consistent with a society in which military discipline and autocratic order are the primary templates. The standard army of conscripted soldiers and daily drill is completely analogous to the present army of industrial workers with their daily programmed response to automatic stimuli. Lewis Mumford notes the effect of the battlefield and arsenal, "Drill made them act as one, discipline made them respond as one, the uniform made them look as one." The drive for uniformity, standardization, and regularity overtook affairs both mechanical and human. Production itself became geared to the demands of military consumption. Today, as well, industrial production is geared first to demands generated by military spending, and since World War II, military research and development.

With the permanent military "state of crisis" fostered by World War II, the intimacy between military and industrial logic achieved an unprecedented synthesis. The development of the atomic bomb forever crippled science in the service of power, and ever since a routine state of war — fueled by the hysteria of the Cold War — has dominated the ideology and economy of American society. One consequence has been that larger and larger numbers of people have become dependent upon the centrally planned mode of production and its means of distribution.

Today over 100,000 companies receive part or whole of their income from the Department of Defense or the Pentagon.[1] The dissemination of social welfare-warfare services by the state entails increasing numbers of people dependent upon centralized power and authority. Mumford illustrates that the "pattern of the new industrial order appeared upon the parade ground and in the battlefield before it entered, full fledged, into the factory." This power nexus begins with the mode of production and distribution of commodities, and reaches out into virtually every aspect of our daily life — from the regulation of time and the workday to the resemblance of the standard military uniform to that of the standard business uniform of dark suit and white shirt. Education, transportation, and many forms of manufacturing derive their basic configuration from military spending.

With the increasing standardization of the industrial machine came increasing magnitudes of power. Economies of scale are translated into bureaucracies of power. The translation of human functions into uniform orderly commodities, into money or units of energy, has its proper expression in the replaceable human units on the assembly line. Undistinguished from the uniformity of the goods they produce, individuals so isolated greatly extend the domination and control that industrial organization extends over other aspects of societal life. What is valued is the machine, what is sought is the most efficient orchestration of the cogs. The army becomes the war machine, and war, the "health" of the capitalist state. The contradiction of this situation should be obvious. War is the ultimate pollutant.

Military regimen is thus the perfect example for increased industrial efficiency. Military requirements have been the major impetus for many aspects of industry and commerce. Military requirements for communication were the largest spur to the communications industry during and after World War I. In recent times, the greatest incentive to the electronics and aviation industries has been military contracting.

The converse is equally true. Not only is military spending a spur to ecologically damaging industry, but the petroleum industry has played a decisive part in the propagation of war for profits throughout the twentieth century. Harold Le C. Ickes, Secretary of the Interior and petroleum administrator for the war in 1944 stated, "Tell me the sort of agreement that the United Nations will reach with respect to the world's petroleum resources when the war is over, and I will undertake to analyze the durability of the peace that is to come."

The history of warfare for seventy years is laced in oil. During World War I, Poland, Rumania, and the Baku fields of Russia were primarily battlefields because of the oil deposits. When the new Communist state accepted the German peace terms at Brest Litovsk, they nationalized the foreign petroleum assets and repudiated all foreign debts. As a consequence, Jersey Standard and Royal Dutch/Shell were the major impetus behind the American and British opposition to the Bolsheviks until 1924.

The petroleum industry extended extensive aid to Japan during the 1930's, and to Mussolini. Disclosures made during the Nuremberg War Crimes Trials and U.S. congressional hearings indicate that Standard Oil of New Jersey had extensive dealings with Nazi Germany's I.G. Farben chemical monopoly both prior to and during World War II. This incident is worth relating both for what it reveals about the intimacy of business and war, and how each provides an impetus for the other.

In 1926, I.G. Farben discovered a process for making liquid fuel from coal. In order to rebuild the German war machine, two joint companies were formed between Farben and Standard Oil of New Jersey which allowed Germany to greatly increase the production of synthetic fuel — a critical need of the Wehrmacht. In 1938, Germany acquired the rights to tetraethyl lead from the Ethyl Export Corporation, owned by Jersey Standard and General Motors. In the late thirties numerous tetraethyl plants were constructed in Germany through a joint venture between Farben, Jersey Standard, and General Motors. A German subsidiary of Jersey Standard designed facilities for high-grade aviation fuel in 1939, at the same time that Jersey Standard extended other valuable military patents to Japan. Concurrently, Jersey Standard acquired German patents for synthetic rubber processing and created a monopoly within the United States.

As late as October 1941, Jersey Standard was involved in court actions to prevent other firms from producing synthetic rubber which was critical to the war effort. Subsequently the government brought Federal criminal conspiracy charges against Jersey Standard. They were denied outside court, but never in legal action. Jersey Standard accepted modest fines and agreed to release its synthetic rubber patents for the duration of the war. Oil, it would seem, is thicker than blood.

ii

The rush to oblivion has dominated the economic activity of the Western world for the last twenty years. In 1968 alone, the world spent nearly $173 billion for military purposes, an amount estimated by Gunnar Myrdal to equal the sum of the total national incomes of all the underdeveloped countries. Of this amount, the United States spent $80 billion and the Soviet Union $40 billion. Today somewhere on the order of 10 percent of the world's output is directed toward military operations employing over 50 million people.

In *The Economy of Death,* Richard J. Barnet estimates that since World War II, the United States alone has spent one trillion dollars on "national defense." This dwarfs all other expenditures regarding the health, education, and welfare of the American people. Between the years of 1961 and 1969, the Federal government spent $10.3 billion on urban problems, $39.9 billion on agriculture, and $507.2 billion on defense.

The ecological and social implications of this phenomenon go far beyond the liberal explanations of the "military-industrial complex." Martin Gellen has advanced the notion that there is no military-industrial complex.[2] The extensive network of military and paramilitary installations America maintains "at home" and abroad are intimately involved in sustaining the extensive economic and cultural investment America has throughout the world. No single sector of the domestic or international economy is dependent upon military spending, but all aspects of production and consumption.

The empire described in Chapter Three involves a reciprocal

relationship between commercial and military affairs. The function of the present military posture of the United States is without question to stabilize the political and economic climate for American investment. On the other hand, military installations both create and act as markets for American production around the globe.

The magnitude of American military operations is staggering. There are some 429 major American military bases around the world with another 2,972 minor bases in some 30 countries covering over 4,000 square miles. The annual cost of sustaining this physical network amounts to over $5 billion annually.

In the armed forces there are almost 3.5 million soldiers, sailors, airmen, and marines with another 1.3 million civilians working directly for the Department of Defense. The Pentagon alone has a monthly employment turnover of 20,000 persons. The purchase value of all Defense Department property holdings, including land, buildings, ships, and airplanes, at cost value, amounts to over $40 billion. Today the value is many, many times that amount. Real property at purchase value totals over $162 billion with personal property amounting to over four times that sum. The influence of this leviathan upon the internal economy of America is well documented. Aside from the direct relationship between the war in Indochina and inflation, there are the remarkable cost overruns for which the Pentagon has become so famous — a strategy now employed by many other systems development agencies. Ninety percent of the time weapons systems run at least double the original estimate. Often such an overrun may present a cost as much as 300 to 700 percent above the original estimate. The case of the C-5A and the persistence of Senator William Proxmire, Democrat of Wisconsin, is a reflection of the situation. In *Report from Wasteland,* Proxmire notes that "Of thirteen aircraft and missile systems with electronic equipment begun since 1955, only four, costing $5 billion, were as much as 75 percent effective. Five more, costing $13 billion, broke down at a performance level 75 percent or less of that fixed in the contract."

This is only one example of the incredible waste and irrationality of military expenditures. In 1969 the Pentagon spent some $38.8 billion on prime defense contracts of $10,000 or more, of this one hundred companies received 68.2 percent, the top ten received 32 percent, and the top five contractors over 20 percent. Military public relations, "the selling of the Pentagon," amounts to nearly $50 million a year.

In short, the planetary and extraplanetary network of American military operations is unparalleled. The extensive system of bases, transportation, communications and weapons systems, both open and clandestine, is of a magnitude that dwarfs the governments of most other nations of the world. The ecological implications of this network are both diverse and profound. The installation of a military base directly influences local ecosystems in innumerable ways. The colonization of Okinawa, the effect upon the Arctic ecosystem and Eskimo culture of the Distant Early Warning Line, isolated radar stations around the world to monitor all variety of activities, supersonic jets — all have deep ecological implications.

The existence of a massive military installation in the Philippines is directly responsible for a huge prostitution business, which profoundly affects the social life of the Philippines. The effects of such a large military installation — elimination of green space, increased sewage disposal, noise, alteration of food patterns, introduction of unique diseases — are numerous. These effects are but suggestive of the consequences for virtually all military installations around the world. Recreation areas are blighted through congestion from servicemen on vacation, surrounding areas are contaminated from the bases themselves, entire national economies become dependent upon the capital flow from American military installations. In the summer of 1966, resulting directly from military involvements, branch banks of both the Chase Manhattan and the Bank of America opened in Saigon. They were built for war: glass blocks instead of windows, and walls designed to withstand mine explosions. *Business Week* notes that "if it weren't for the massive U.S. presence there, probably neither bank would be in Vietnam."

This is just one illustration of the intimate relationship between American military activity and the investment of American corporate capital, or, the cooperation between the state and corporate sector. As we shall see later in regard to petroleum, this occurs on a regular basis throughout the world. The political composition of many nations is a direct result of American military intervention — and not just in Vietnam: "For more than twenty years, the United States has carried on a global campaign against revolution and native insurgency movements, conducting a major military campaign or CIA operation in an underdeveloped country about once in every eighteen months."[3]

In lieu of direct intervention, terror is a handy tool. The ecological consequences of weapons development, testing, and use have important implications. The results of atomic warfare upon Japan — twenty-five years ago — are still evident in the landscape and social fabric of that nation as Robert Jay Lifton has documented in *Death in Life*. The psychological terror the world endures because of such a threat has had a deep and unalterable impact upon the human psyche — enough so that an entire generation is identified as the "Hiroshima Generation."

Nuclear power in Ralph Lapp's words is the "supreme and acute pollutant of our times." The biological and social impact of nuclear and hydrogen weapons testing has had a pronounced influence upon the natural environment. We have not hesitated to eliminate entire islands for the sake of weapons testing. Bikini Atoll and Eniwetok in the Pacific, names to us, but home to people of that region, are now uninhabitable as a result of testing. Vast stretches of the Nevada desert are forever uninhabitable. The incidents of atomic accidents are by now well documented. Several planes carrying nuclear explosives have crashed: 1961 over Goldsboro, North Carolina; 1966 over Spain; 1968 over Greenland. These are only three of some twenty-two nuclear accidents — called "Broken Arrows" by the Pentagon — to which they admit only thirteen. To this estimate we must add the contamination of the earth's soils and water from strontium 90 and iodine 131.

Besides such direct contamination, the United States has dumped nuclear "refuse" into the oceans on a regular basis for the past twenty-five years. This occurred for well over a decade before the Atomic Energy Commission even told the city of San Francisco what they were doing offshore. The Atomic Energy Commission now has on hand some 80 million gallons of nuclear waste which they plan to dump into a mine shaft in Kansas. Annually the AEC produces 2.4 million cubic feet of nuclear garbage.

Compared to the research and investigation directed toward producing such hazardous materials, little concern has been shown about the consequences of such practice upon people or other living organisms. When such speculation occurs, the AEC all but ignores it, or seeks to discredit the source. A University of Pittsburgh radiologist, Ernst J. Sternglass, has suggested that the infant mortality rate in the five years following the first small atomic blast in New Mexico in 1945 went up by 40 to 50 percent in the path of the fallout cloud which included parts of Texas, Arkansas, Louisiana, Mississippi, and the Carolinas. He also believes that records from the West Coast suggest similar findings subsequent to the Pacific Bikini tests.

Sternglass offers similar observations regarding the effects of gaseous emissions from nuclear power stations. Appearing before a Pennsylvania State panel investigating nuclear power plants in the state, Sternglass suggested that the Dresden 1 Nuclear Station of Commonwealth Edison, fifty miles southwest of Chicago, was responsible for up to 2,500 extra infant deaths in the state of Illinois between 1959 and 1969 from the gaseous emissions. This finding, as with the one regarding infant mortality on the West Coast, has been widely questioned. But more important than the actual accuracy of the Sternglass analysis is what it reveals about our ignorance regarding the effect of nuclear activity over the past twenty-five years. In 1969 Sternglass suggested that 400,000 possible infant and fetal deaths had occurred in the United States as a consequence of the fallout from weapons tests in the 1950's. Arthur Tamplin, who had been studying the effects of radiation upon man and the environment at the AEC's Livermore Laboratory in California, was given the assignment of critically analyzing the Sternglass findings. Tamplin's original estimate of 4,000 deaths was considerably below that of Sternglass's. Together with Dr. John Gofman, he suggested that Federal radiation standards for acceptable radiation were ten times higher than they ought to be. Failure to tighten standards, they thought, would result in 32,000 additional cancer and leukemia deaths. In March 1971, their estimate rose to 74,000 deaths. Since these findings Tamplin and Gofman have received constant harassment from the AEC. Their book, *Population Control Through Nuclear Pollution,* documents their findings and the story of their struggle to question the acceptable standards of the AEC. Such standards are made on the assumption, as Representative Chet Holifield (Democrat of California), chairman of the Joint Committee on Atomic Energy has said, that "precautions are being taken by those who know."

Given the past performance in regard to the premature deaths of pilots who flew test planes over the atomic tests sites, the premature deaths of people who live in the towns around the Nevada test sites, the documented evidence linking uranium miners to a high incidence of lung cancer, the thyroid disorders just now developing in Marshall Islanders exposed to unexpected radiation from fallout, et cetera, one is reminded of a statement by Clement Attlee, who, when asked about fallout from the first atomic bomb said, "I knew absolutely nothing about the consequences of dropping the bomb, except perhaps that it was larger. We knew nothing whatever about the genetic effects of an atomic explosion. I knew nothing about fallout. As far as I know, President Truman and Winston Churchill knew nothing of these things either."

Now, it would seem, we do know something. Does that make a difference? To supplement the supply of nuclear weapons, the AEC now operates 17 nuclear reactors; has 54 under construction, 38 more in the final planning stages, and at least another 9 plants scheduled for construction. Within 30 years the AEC will probably have 950 licensed power plants around the country. The potential hazards from such plants are of three basic kinds: catastrophic accident or what is technically known as "nuclear excursion" or a "core meltdown"; thermal pollution, the effluent discharges which occur in the routine operation of nuclear reactors and fuel-processing plants; and the task of disposing of the nuclear wastes. The safety of such operations is the subject of a wide-ranging controversy in the United States, of which Gofman and Tamplin are but the tip of an iceberg.

At present one-half of the United States research and development budget goes into arms development. Somewhere on the order of 53 percent of the AEC budget is allocated for military purposes. And yet Dr. Paul Tompkins, the executive director of the Federal Radiation Council has stated in regard to the Gofman-Tamplin tenfold radiation reduction recommendation:

> To reduce radiation exposure tenfold would cost billions; it might even cost more than the Vietnam War. To comply, you'd practically rebuild all nuclear installations and the factories that use any sort of X-ray equipment. We'd have to review radiation exposures for wristwatches, TV sets, and radium dials. Plus, I'm not completely sure it is not technically possible to monitor down to such a tight level.[4]

What is at stake with the development and use of nuclear power is not the intentions of the men involved or the nature of atomic knowledge itself, but the society in which such knowledge is created. The men who originally advocated nuclear power, such as Albert Einstein or Leo Szilard, were humane and sensitive people. But not sensitive enough:

> The penalty for producing nuclear bombs sufficient to destroy the human race was that it put those genocidal and suicidal weapons in the hands of demonstrably fallible human beings, whose astounding scientific achievements blinded their contemporaries to the human limitations of the culture that had produced them.[5]

At present numerous new scientific accomplishments, from the mastery of chemical and biological agents to the creation of "life" in a test tube may well suffer the same use as the atomic discoveries. James Watson, the discoverer of DNA and the author of *The Double Helix*, the story of his discovery, has warned as much about the imminence of genetic engineering, though science blunders along in its so-called objectivity while events of cosmic magnitude are developed in the laboratory at government expense.

The corruption of scientific development comes from the coalition of science with power to the extent that scientific thought has become almost wholly dependent upon government and corporate support. Science in the Western world, rooted in the man-centered, mechanistic attitudes we described earlier, increasingly fostered the progressive reduction of the dimensions of life — the fragmentation, orderliness, routine, and monotony which today typify our society. The mass production of commodities goes hand in hand with the mass production of knowledge. With this increasing fragmentation the sense of limits, of rational contours which comes naturally with the perception of the whole, gradually disappears. The dictums of efficiency and progress replace the more "dated" organic conceptions of science. There are innumerable examples of the hazards of weapons development and testing and their ecological implications for the scientific enterprise, and society as a whole. One other example, however, should be considered which, from an ecological viewpoint, is perhaps most significant of all — the development of chemical and biological weapons. While the threat of nuclear holocaust is a "clear and present danger," these other weapons systems with the same magnitude of destruction have been quietly developed, tested, and employed.

While several works now exist which begin to document the dangers involved, the development of such systems is scantily understood. While public opinion has accepted the myth that such weapons have been developed and tested solely as defensive weapons, we know the opposite to be the case. One such agent, GB, derived from Nazi research on Sarin during World War II, was *first* manufactured by the United States during the Korean War as an offensive weapon against Communist China. Absolutely no evidence exists to suggest that the Chinese were either developing or even considering the development of such a nerve gas. This and other examples reveal the incredible public mythology the government has built up around CBW as strictly defensive weapons, when in fact the United States has pioneered in the initial development. The United States Army, as of January 1970, has trained 550 foreign officers of some 36 countries at Fort McClellan, Alabama, in "defensive aspects" of chemical and biological warfare. The army admits that "it is not possible to separate offensive tactics from defensive [and] there can be no absolute guarantee that defensive tactics will not have some utility in forming offensive tactics."

To date, the United States has instructed military personnel in CBW warfare from such nations as South Vietnam, South Korea, Thailand, and several Arab nations. Representative Robert W. Kastenmeier, a Wisconsin Democrat, in December 1969, stated, "It should be noted that the United Arab Republic used poison gas in

Yemen in 1965 and that it was reported that Israeli forces advancing into Sinai during the June 1967 war uncovered chemical warfare equipment." The almost routine "accidents" involving the testing of CBW agents are well known. Open-air testing of nerve gas occurs regularly at the Dugway proving ground in Utah, Edgewood Arsenal in Maryland and Fort McClellan in Alabama. At Dugway, in March 1968, 6,400 sheep were killed by a "change in the wind." Spillage of CBW agents has occurred several times and caused both evacuation and pathological consequences to people at Dugway as well as on Okinawa. Somewhere between 700 and 800 accidental laboratory infections have occurred at Fort Dietrick and Pine Bluff Arsenal between 1954 and 1962. In August 1970 the army, in an unusually publicized event, dropped canisters of nerve gas off the Florida coast in 16,000 feet of water, without conducting a single test to determine what the effect of the pressure would be upon the canisters.

There is no end to the horror of American weapons development and testing. There is no way to eliminate that horror except to eliminate the development of such agents altogether. The notion of "permissible risk" in the use of atomic and CBW agents is ecologically unacceptable yet an everyday practice — in the face of repeated "accidents." Here is but one more example of the manner in which the structure of military development and employment counters fundamentally the requirements of an ecologically sound society. "Permissible risk" entails the estimate of tolerance levels, both in regard to the amount of danger that can be safely absorbed, as well as the odds of miscalculation. Safety regulations are then established in regard to the amount of "dangerous" material that can be "safely" absorbed. This, of course, occurs without the advantage of thorough and long-term testing. This procedure completely ignores the realization that even if the acceptable level is not lethal today, the accumulation resulting from the calculated tolerance will, over time, become severe — it may become terminal. The recent incidence of men contracting cancer some twenty-five years after working with the AEC during the first atomic experiments is symptomatic. The entire "fail safe" notion adapted from the military is today prevalent in regard to chemicals and other agents employed in the foods we consume, the tolerance levels established for appliances, for carbon monoxide and other gases emitted from the internal combustion engine, and a host of other everyday occurrences. We are not accustomed to grasping the complexity of the whole picture. We would rather consider a "newsworthy" nerve gas spill or a bad day for air pollution in New York City. As a result, we rarely consider the total cost, ecologically and socially, of what militarism has meant over the past fifty years. We assume, rather matter of factly, that "life will go on." In order to dispel that illusion, we might consider the ecological warfare that has been waged against the people and landscape of Indochina.

iii

In *Ecocide in Indochina: The Ecology of War* I have documented in detail the ecology of war in Indochina.[6] Here in microcosm is a picture of the magnitude of the war establishment as it affects the culture and landscape of a region. One cannot consider the ecological politics of the United States without considering what has been wreaked upon the peoples and landscape of Indochina.

In Indochina there is no mushroom cloud or stench from the gas chambers. There is no clear image or date by which to mark or measure what has happened there — what continues to happen. Yet in Laos alone, during 1970, the equivalent of several Hiroshima devices were dropped in "conventional" weapons every month. While the weapons of saturation bombing and chemical warfare seem to be of a more limited nature than Hiroshima or Auschwitz, they are every bit as profound in their impact. Studies conducted by North Vietnam reveal that defoliants "provide important chromosomic alterations in the local population."

In South Vietnam between 1961 and 1970, the U.S. Government has applied six pounds of herbicides for every inhabitant, or twenty-seven pounds per acre. The National Institutes of Health found as early as 1965 that 2-, 4-, and 5-T, used in 90 percent of the missions, was teratogenic (fetus deforming), but did not release the information until 1969, and then under pressure. On December 30, 1970, the army revealed it knew that defoliation was destroying mainly civilian crops as early as 1967, but decided to continue spraying. American estimates suggest that between 15-20 percent of the forest areas and 7 percent of the arable land have been sprayed with herbicides. Figures for the Democratic Republic of Vietnam indicate that 44 percent of the forest areas and 43 percent of the arable land have been sprayed.

By June 1970, in South Vietnam alone, the government dropped enough ordnance to amount to roughly 1,200 pounds for every man, woman and child, and another 130 pounds for every square mile. The 3.5 million bomb craters in South Vietnam constitute the largest excavation in human history — ten times that of the Suez and Panama Canals combined. The 2.5 billion cubic yards of earth would cover the surface area of the state of Connecticut.

It is impossible to talk rationally about ecocide in terms of limited warfare. The ascribed limits of national conflict do not apply to the overarching natural interface of environmental systems. With ecological warfare, national confrontations become issues of global survival.

Chemicals dispersed over South Vietnam find their ecological path into Cambodia and throughout the Mekong River Basin. Some persist in the soils for decades. To understand ecocide requires a breadth of imagination and scope of understanding which is not now present in our conception of the second war in Indochina. It is simply not enough to consider news events in isolation. To understand the destruction of Indochina or America requires that we come to understand the ecosystem of Southeast Asia as one single organic fabric in which all living things are tied together by an infinite number of interdependent strands. Obviously we have not been able to grasp this about the United States, let alone Indochina.

The almost total disregard for the ecological implications of the war in Southeast Asia is well demonstrated by the fact that the first on-the-spot government inspection of defoliation did not occur until 1968, seven years after the program began (Operation Ranchhand — "Only We Can Prevent Forests"). The first nongovernmental inspection of the ecological damage occurred in March 1969 by Gordon Orians and E. W. Pfeiffer. During the summer of 1970, a study team of the American Association for the Advancement of Science examined the damage and reported back to the AAAS convention in December. A resolution was passed, and Glenn Seaborg, chairman of the AEC, was elected president of the AAAS. The extreme irrationality of the entire program has been demonstrated well by Thomas Whiteside:

> It can be estimated that the American military destroyed the rice crop of a million people with the aim of denying food to twenty thousand Vietcong. Or, to put it another way, in order to deprive the Vietcong of one ton of rice the American military has to destroy fifty tons of rice that would ordinarily support members of the civilian population.[7]

The advantages of chemical warfare in Indochina and against all guerrilla activity are becoming increasingly apparent to U.S. militarists. Increasingly herbicides will be used against the civilian population as a means of forced urbanization. The destruction of foliage and crops forces the peasant population off the land and into the cities. This can be accomplished without the public attention afforded large military operations or tactical air strikes. The effect is more permanent than bombing. The rural culture is destroyed biologically without opposition. Chemical warfare thus allows the U.S. Government to minimize the dependence upon troops and air units while conducting a war with old C-123 cargo planes filled with chemicals.

The long range implications of this warfare are staggering to contemplate. We are not concerned simply with the destruction of crops and foliage, but the poisoning of people, birth defects resulting from chemicals, forced urbanization, the destruction of rural culture, and the obliteration of the symbiotic relationship between the Indochinese people and the natural environment. This in turn affects the health of the population, the transmission of culture, the role of men and women in the society, and the intellectual growth of children. As one Vietnamese put it, "Long after the U.S. withdraws her troops and technology, America will remain in Asia in the soils, air, water, biological fabric, and agony of her scarred people."

On August 26, 1970, the United States Senate voted 62-22 *against* banning the use of herbicides in Indochina. The United States continues to set in motion biological mechanisms which will be at work in Indochina for decades, if not longer. There will be no postbellum reconstruction of Southeast Asia. No Mekong TVA can restore the nutrient value of the soils of Indochina. No Asian Marshall Plan can restore the natural growth of a generation of children starved into intellectual retardation. What kind of arrogance will be required to send our teams of scientists and social engineers into Indochina to "patch-up" the damage, to study the Vietnamese as we did the Japanese after Hiroshima?

iv

This vast interconnection between American military and ecological policy is far from accidental, but rather results from a single uniform policy implemented systematically over many years. No other person in the Federal government demonstrates that connection as well as the "Senator from Boeing," Henry S. (Scoop) Jackson, a Democrat from the state of Washington, the chairman of the Senate Committee on the Interior and Insular Affairs, and ranking member of the Senate Armed Forces Committee as well as the Senate Appropriations Committee. For over a decade, Jackson has been a key figure in the Democratic Party hierarchy, nominating John F. Kennedy to the presidency in 1960, and a potential candidate himself in 1972.

Just a brief look at the career of Henry Jackson is quite revealing. No other person in the U.S. Senate has been more instrumental in the development of U.S. air superiority than Senator Jackson. Coming from the state of Washington, home of Boeing, this is not hard to understand. But there is more to Jackson's fervent desire for military hegemony than simple economics. He has been instrumental in promoting the theory of nuclear preparedness. Together with his nuclear and air strategy, he has systematically promoted policies which can be described as nothing short of a complete contradiction of his post as chairman of the Senate Interior Committee. Jackson has been perhaps instrumental in the preservation of more than one natural wildlife refuge, but at the same time, more than any other man in the Senate, he has laid the foundation for the destruction not only of the wilderness, but of the entire earth. The up-front contradictions between the claims of his ecology constituency and his role as the chief Senate hawk are not without interest. His failure to stand against the testing of nuclear weapons on Amchitka Island in Alaska, despite the loud cries of ardent conservationists, stands out. Not to be swayed from wooing Senator Jackson into more parks, the Sierra Club gave the senator their annual John Muir Award in 1969, obviously more concerned about parks than Asians.

If the senator's performance has been less than the mirror of integrity, his behavior when the claims of the earth and of the military clash has been something less than noble. Take for example his role in exempting the military from construction of waste water treatment facilities. Senator Jackson, in regard to Section 808 of the Military Construction Authorization Act for fiscal 1968, defended provisions which essentially prohibited the Department of Defense from the construction of waste water treatment facilities that would be more stringent than the local community's. The net effect being that in many communities, where no such facilities, or facilities which provide little if any actual treatment, exist, the military would be prohibited from improving on the conditions of the local community. As part of the colloquy, Senator Jackson explained that "I just wish to make the general comment that what we were really trying to move against was the idea of making the Department of Defense the guinea pig in communities and sections of the country where there had been no real effort to move on pollution abatement." Of course one might be somewhat sympathetic to the senator, since for the Department of Defense to move against pollution would in fact be to move not just for water treatment plants but for the abolition of the department itself. Nothing less than the abolition of the military will curtail the hazards outlined in this chapter. This is not a matter of monitoring or cutting down militarism in American society — it makes little difference if one or fifty nuclear bombs are dropped on New York. One is enough. The inability of the constituted congressional procedures to check the escalation of the war in Indochina, first into Cambodia and then into Laos, demonstrates the fact that only disarmament will curtail the use of arms, not good intentions or new legislation.

Such a task requires much more of an understanding of the

current military juggernaut than presently exists. Militarism constitutes 10 percent of the American GNP. We are not simply talking about selected aircraft companies such as Lockheed or Boeing, but about corporations which are at the very center of American production and capital formation; AT&T, General Motors, Ford, IBM, Standard Oil of California. Almost every major war contractor is also involved in nonmilitary forms of production, such as transportation or educational equipment.

Over a hundred major firms whose traditional manufacturing activities lie outside the defense sector have diversified into it since the end of the fifties. These companies include every major sector of American industry including automobiles, petroleum, nonelectrical machinery, metals, textiles, rubber, electronics, and precision instruments, to note just a few. Harry Magdoff has summarized some of the implications of the military establishment for business:

> The widespread military bases, the far-flung military activities, and the accompanying complex of expenditures at home and abroad serve many purposes of special interest to the business community: (1) protecting present and future source of raw materials; (2) safeguarding foreign markets and foreign investments; (3) conserving commercial sea and air routes; (4) preserving spheres of influence where the United States business gets a competitive edge for investment and trade; (5) creating new foreign customers and investment opportunities via foreign military and economic aid; and, more generally, (6) maintaining the structure of world capitalist markets not only directly for the United States but also for its junior partners among the industrialized nations, countries in which the United States business is becoming ever more closely enmeshed.[8]

Military spending has, since World War II, become the major support of the capitalist state. Martin Gellen notes that at home, the large-scale spending by the military is a classical Keynesian technique to stimulate private investment in goods and services, seeking to increase equilibrium through steady employment and income. However, such spending invariably tends to bolster the consolidation of the economy in the hands of the private sector as well as socialize the private costs of producing commodities which can have no direct use value to the general public. Spending for programs or services which the public authentically requires, health, education, and welfare, is minimized. That is to say that the "military-industrial complex" is merely a system by which to maintain a system of public exploitation for private gain. Military spending functions first and foremost to maintain the profit structure of private capital. It is in no way necessary to maintain the present level of life in America; it is, however, critical to sustain the structure of capital formation. This form of accumulation, in which the rich get richer and the poor remain about the same, is masked by the investment and employment that military spending provides. Stagnation thus comes to the forefront as military spending diminishes, such as the present when minimal withdrawals are occurring from Indochina. Fortunately for the American economy, World War II provided a mask of employment and spending to hide the critical economic conditions and Vietnam postponed the reality. There is no escaping the contradictions and irrationalities of capitalism. Perhaps the ultimate contradiction resides in the fact that the more society spends for arms, the more likely will be the destruction of that society.

Today with military spending down, major war contractors are diversifying their operations into other markets in order to make up for the fluctuations of defense spending. Their position of hegemony within the economy, in most cases, remains unscathed.

It is obvious then that military spending is not simply a matter of altering priorities, as Senator Proxmire would do by merely requiring firms receiving war contracts to diversify into other forms of production. It is in this context that the proposals for "reconversion" or "conversion" must be understood. To begin with, they are quite different, though often confused. "Reconversion" implies a return to a prior state, something along the lines of the minimal reconversion which occurred subsequent to World War II. To talk today about conversion is part of the larger public concern for "new priorities," or shifting around present expenditure from war-related enterprise to more useful programs and services. In the summer of 1969 Senator Abraham Ribicoff initiated a questionnaire sent to 118 corporations and 18 major cities requesting their views on economic conversion. The response was that,

> Most industries have no plans or projects designed to apply their resources to civilian problems. Furthermore, they indicated an unwillingness to initiate such actions without a firm commitment from the government that their efforts will quickly reap the financial rewards to which they are accustomed. Otherwise, they appear eager to pursue greater defense contracts or stick to proven commercial products within the private sector.

But one might well respond by wondering whether or not it is desirable, for instance, for Boeing simply to shift from the production of military aircraft to mass transit in the city of Seattle? Is the economic and political power of Boeing over the economy of Seattle and the Northwest any less insidious? Obviously not. Even if several large war contractors were to shift the focus of production from war-related to other kinds of production, their position of political and economic power within the society would remain unaltered, and instead of producing chemicals for the destruction of Indochina they would merely continue to produce chemicals to destroy America, as before — though with less visible consequences than in Indochina. Martin Gellen has captured the essence of what is involved in the plans for economic conversion: "Plans for conversion totally avoid the questions of how and by whom such allocations have been made in our society and are to be made in the future. In short, they avoid the question of the lack of popular control over the public sector of the American economy."[9]

To sum up, any meaningful attempt to alter the present pattern of military and paramilitary activity by the United States can only succeed if it challenges the foundation of the power behind that activity, the power which enables the present government to govern, not merely seek to alter the use and management of that power. That entails nothing less than disarming the United States Government as it presently exists. Present military expenditures are in no way necessary to maintain the present level of welfare of American society, they are required, however, to maintain the present distribution of wealth within that society.

The severe forms of social and ecological imbalance that result from the existence of weapons and other military systems is without question a daily threat to the well being of this entire planet. There can be no ecological or social balancc in this world without the elimination of the military impetus for the organization of society. There are basically two paths available. Either preclude the rather naive assumption on the part of some that the present power of the state will simply hand over control or change their consciousness. One is a decision on the part of the ruled that they will no longer recognize the authority of the ruler. Namely, that the Department of Defense is deprived of an army. While opposition within the military is a very encouraging development, plans for a volunteer army may well curb any real possibilities for undermining military manpower requirements.

Second, the citizens of the United States could refuse to continue to financially support militarism. To stop paying Federal or state taxes. What this would entail is the development of authentic and unique centers of political power within the United States which gradually organized a framework of political and economic autonomy. Such a notion entails the questioning of the present form of government and the scale of political and industrial organization. It would entail local political organizations which dared to divert Federal taxes into local political organization. The city of San Francisco along pays annually to the government in taxes a sum comparable to its entire annual budget. There is no other way to meet the financial crisis of the urban areas. There is no other way, as well, to thwart the American military.

At the center of the future is the struggle for political liberation, nothing less.

V

Oiling the Machine: Automobiles and Petroleum

Automobiles and petroleum constitute the most dominant influence upon, and the most cohesive element of the American international economy.[1] No other factor has exercised such a predominant influence upon the direction and definition of the American industrial empire. In terms of capital formation, patterns of employment, the extent of marketing penetration, transportation, energy, urbanization, ecological imbalance and American foreign policy, they are unrivaled in influence. The impact of their economic priorities upon the public policy of the United States is unsurpassed by any other lobby or combination of lobbies. The network of subsidiaries, affiliates, and joint ventures runs into the tens of thousands and affects the political economy of virtually every nation. Automobiles and petroleum determine the political composition of governments and the chemical composition of the atmosphere.

One way to readily grasp the magnitude of what is involved is a comparison of the ten non-auto or oil corporations among the top twenty U.S. industrials and those ten *directly* related to oil and autos:

		(In Billions)	
	Sales	*Assets*	*Net Income*
Non-oil or auto	\$48.2	\$47.1	\$2.7
Oil and auto	\$89.5	\$86.3	\$6.2

SOURCE: "The 500 Largest Industrial Corporations," *Fortune* (May 1970).

Internationally, nine of the top twenty-five industrials are directly petroleum or automotive industries. These figures do not take into account the fact that among the top twenty U.S. industrials, U.S. Steel and Goodyear Tire and Rubber are directly dependent upon automobile manufacturing for the bulk of their market. Moreover, General Motors is the offspring of Du Pont, the source of many GM automotive "improvements," such as lead for gasoline, paints, and other products.

Yet only within recent years has public attention become focused, albeit slowly, upon the automotive and petroleum industries. This attention first arose in regard to problems of consumer safety with the automobile and more recently in regard to the costs of air and water pollution from oil and the internal combustion engine. Today these industries are the target of a wide variety of public struggles against air pollution, oil spills, pesticides, the Alaskan pipeline, nondisposable plastics, land-use patterns, mass transit and the monopolization of energy supplies. While the role of petroleum investment in American foreign policy has received some consideration, there exists not a single definitive work on either the role of oil in the affairs of the United States in the Middle East or its impact upon the course of the war in Indochina. Less recognized is the overarching impact of oil and automobiles upon the entire fabric of American foreign policy as it affects the Third World, Western Europe, and Asia.

Least understood is the tremendous impact the development of the automotive-petroleum empires has had upon the culture and economy of the United States as world power. The common perception is that petroleum companies produce gasoline for cars and automotive companies merely produce automobiles. This is a very crude beginning of an understanding of a force within American society which may constitute as much as 40 percent of the entire GNP. Automobiles and petroleum together constitute nothing less than a microcosm of America itself. They operate in more nations than the Department of Defense with assets and expenditures which annually dwarf defense spending. Boasted Robert Stevenson, Ford executive vice-president for International Affairs:

> We at Ford Motor Car Company can look at a map without any boundaries. . . .
> We do not consider ourselves basically an American company. We are a multinational company. And when we approach a government that doesn't like the U.S. we always say, "Who do you like? Britain? Germany?" We carry a lot of flags.

ii

Since the turn of the century the manufacture, distribution, maintenance, and commercial use of motor vehicles have been the glue of the American economy. In the first ten years of the century the automotive industry hardly existed at all. By 1925 it had achieved an unrivaled leadership in the value of products, cost of materials, and wages paid. Between 1919 and 1941 the annual production of motor vehicles increased more than two and a half times, the real value of output, three times. This was due in large measure to the introduction of the mass assembly line by Henry Ford and the development of innovative financial mechanisms along with the extension of consumer credit which characterized the ascendency of General Motors. These two developments became the cornerstones not only of motor vehicle production but all forms of commodity production. As such, automobiles set the pace for American economic activity from the 1920's through the 1960's. While railroads were the center of the American economy in the second half of the nineteenth century, automobiles were the primary source of economic development in the first half of the twentieth century.

Ford introduced the moving assembly line, the multipurpose precision machine tool, the branch assembly plant, and a host of other production techniques. If Ford provided the possibility of a car in every garage, GM made it a reality. General Motors introduced the extensive product line, the annual model change, massive advertising, the "trade-in," and most important, consumer financing and the tools of systematic market analysis. Consumer credit first appeared with the auto and has been with us ever since.

The credit card boom was pioneered by the gasoline companies and was a natural counterpart to the auto credit system. In 1969, automobile credit amounted to $37 billion.

The arrival of the horseless carriage had the most profound effect upon petroleum. By 1913 there were one million motor vehicles on the roads, in less than ten years the number had grown to 10 million. It has been nonstop ever since, and so has the flow of oil. In 1900 the United States consumed 40 million barrels of oil. That was 4 percent of the national energy budget. By 1929 the United States was consuming one billion barrels per year, up to 2 billion in 1950 and 3 billion in 1962. Given the ravenous escallation, it seems hard to explain why the present destruction of the atmosphere or congestion of the highways was not anticipated years ago. Or was it ignored?

The motor car was a market impetus to more than just the oil companies. Rapid economic expansion and consolidation were motivated in many sectors of the economy by the requirements for mass automobile production and distribution. Motor vehicles created a mass market almost overnight for such staples as steel, rubber, plate glass, aluminum, leather, paint, and a vast variety of other products. In 1968, automotive materials consumption looked like this:

Item	*Percentage of Gross Market Consumption*
Steel	21.0%
Aluminum	10.4%
Lead	54.7%
Nickel	14.3%
Natural rubber	68.8%
Zinc	36.5%

SOURCE: Automobile Manufacturers Association, "Automotive Materials Consumption, 1968," *1970 Automobile Facts and Figures,* p. 34.

The automotive industry concerns more aspects of industrial production than any other economic activity. A small passenger car involves 2,500 major parts and assemblies. In total, counting every nut and bolt, there are over 20,000 parts involved in the assembly of a single auto. The production of a motor vehicle depends upon 250 different chemical products and over 80 pounds of plastics per car. In a single year this accounts for 65 million pounds of cotton, 7 million pounds of nylon, 50 million pounds of vinyl, 12 million pounds of rayon, 24 million square feet of leather and 50 million gallons of paints and lacquers.

However, the dimensions of the industry can only be grasped by realizing that motor vehicle production spends more money for purchases within its own sector than from steel, rubber, glass, plastics, and electrical equipment and tool manufacturing combined. This means that the purchases of automobile companies from their own subsidiaries or other strictly auto-parts manufacturers surpass the purchases of materials in dollar value from outside the industry. In 1967, purchases from within the automotive sector amounted to $14.8 billion, well over four times the purchases from steel. In order to supply the automotive industry, innovations are generated within other industries for motor vehicles. The steel industry developed alloys for lighter metals, the petroleum industry developed high octane gasoline, the glass industry developed shatterproof glass, and new construction techniques were developed to accommodate automobiles and highways.

In the United States alone there are 700,000 registered mechanics, which is one for every 130 autos — a ratio better than physicians per person. This is apart from the 217,000 employed mechanics in 112,000 independent repair shops and the home mechanics who tinker with their automobiles. In total, more than 14 million people, or 16 percent of the work force, are today concerned with the production, distribution, and "care" of the automobile. This represents 13 percent of the entire GNP and more than 822,754 different businesses, and constitutes a remarkable proportion of the entire economy:

	Establish-ments	*Employ-ment*	*Payroll*	*Sales and Receipts*
Manufacturing	1.1%	5.0%	5.9%	11.6%
Wholesale	21.1	15.1	13.6	17.3
Retail	17.4	15.8	19.1	24.4
Selected Services	16.6	15.7	12.5	17.1

SOURCE: *1970 Automobile Facts and Figures,* pp. 36-37; see source in preceding table.

Automobiles represent 24 percent of all retail sales and consume an incredibly high proportion of the nation's personal income. In 1969 estimated personal expenditures for new and used autos were $36 billion out of a total $91 billion spent on all durable goods, or 39.6 percent! The proportion of an individual income spent for automotive purposes is rising. During the years of 1961 through 1969 the cost of motor vehicle accessories increased by 28 percent for motor oil, 31 percent for tires, 41 percent for auto insurance, 28 percent for auto registration fees, and 21 percent for repairs. The production of tires, batteries, other accessories, and the vast development of customized parts, is a major automotive subsidiary. The inflated costs of the motor vehicle industry are an important contributor to the general inflation of American society — a force matched in comparison only by the costs of the war in Indochina.

The nature of automotive production today is in no little part responsible for continued inflationary tendencies within the economy. General Motors alone spends $240 million a year on advertising. There are today 30 different nameplates and 360 different models of the automobile produced in the United States. The proliferation of automobiles in sundry sizes, shapes, and styles has nothing to do with increased use value to the public. Today only 47 percent of new car sales are traditional models, the remaining 53 percent of sales are geared to appeal to particular fancies, ranging from masculinity and sex to economy and foreign-made snobbery. All basic engineering innovations in the automobile, such as automatic transmission and synchromesh gears, were developed prior to 1940. Subsequent developments, including chrome, interior "comfort," and the tail fin, account in large measure for the fact that between 1947 and 1956 the return on stockholders' investments by the Big Three was twice as great as the average for all manufacturing companies. General Motors makes three times the profit on any one vehicle as a foreign competitor.

With tail fins out and consumer concern increasing, the automotive industry has begun to market safety instead of comfort or sex appeal. As one commentator noted, "GM is prepared to derive profits from the sale of any conceivable automotive commodity."[2] In 1968, the Big Four auto manufacturers recalled approximately 30 percent of their entire production line for safety defects. The magnitude of the automotive sham is illustrated not only by the lack of safety, but the basic inefficiency of the technology itself:

> The automobile is a particularly good example of the inefficiency of the energy in crude oil, 13 percent is lost in refining, 3 percent is used in transport to the consumer, 25 percent is converted to work in the engine but only 30 percent of this amount is transmitted to the road (after losses due to friction and auto auxiliaries), and further decreases occur through gears and tires. The overall efficiency of the automobile is about 5 percent, though air drag, braking, and idling reduce this in actual operation.[3]

The influence of the automobile on other aspects of American society, notably unions and transportation, is profound. The first large-scale union of the century appeared in automobile production. The transfer early in the twentieth century from a craft orientation to a concentration on industrial production as the focus of union organization was rooted in the growth of the United Auto Workers. This resulted in large measure from the very high percentage of all employment that was dependent in some part on the automotive industry, as well as on the highly automated forms of mass assembly line production, which had all but eliminated craft and guild positions.

Automobiles dominate the present system of transportation. In 1969, they accounted for 87 percent of all intercity transport, while eight out of ten commuters relied upon the car. Even with the airplane, 86 percent of all long-distance trips are taken in the auto. Today, 80 percent of all American households own at least one car, 29 percent two or more.

As of 1968 the United States produced and assembled 10 of the 28 million cars and commercial vehicles in the world (not counting American-produced vehicles overseas). This figure has slowly declined as American firms extend their overseas production facilities, which do not appear in the official figures for domestic production:

Year	*Production Percentage*	*Consumption Percentage*
1953	70	67
1960	50	62
1965	50	49

SOURCE: Calculated from the *UN Statistical Yearbook,* 1969.

While the domestic United States today constitutes one-third of the world's production of automobiles, in 1969 the United States had more motor vehicles registered than the next twenty-six largest nations combined. There are more motor vehicles registered in Los Angeles County than in all of Africa. However, the American hold on the automotive market is declining steadily. In America, one

Production of Motor Vehicles 1969 (Millions)

	1968	*1965*	*1960*	*1953*
World	28,250	24,310	16,400	10,490
U.S.	10,718	11,056	7,868	7,322

Motor Vehicles in Use (Millions)

World	216,281	178,620	126,320	82,340
U.S.	98,563	89,093	77,896	55,530
Africa	4,090			
N. America	110,130			
Europe	67,380			
Asia	16,380			
S. America	6,170			

SOURCE: *UN Statistical Yearbook,* 1969, p. 21.

auto for every two and one-half persons seems to have reached something like saturation. As a consequence, automotive producers have diversified. Most reimport many components of automobile production, that is, parts are produced in Europe or Asia and shipped to the United States for construction. GM's German-produced Opel is the third largest import in sales in the United States. The nonautomotive production of General Motors includes tanks, rifles, navigation systems for the Apollo, refrigerators, and many other products.

Within the United States, GM accounts for over half the total passenger cars produced, twice the production of Ford and three times that of Chrysler. These three companies produce 83 percent of the nation's autos. Ford has consistently been the most international of the Big Three. Over 31 percent of Ford's total unit sales occur overseas, as compared to 19 percent for GM.

The monopoly of American automotive production worldwide is a microcosm of the position of General Motors at home. Though GM produces most of the components involved in the production of a single automobile, it still relies upon 37,000 outside firms for goods and services. The importance of this vast product market is illustrated dramatically every time there is a strike against GM. All major sectors of the economy are affected.

The magnitude of influence General Motors exercises over the course of the entire economy can be illustrated by dollar comparisons with other corporations in 1969.

Company	*Sales Billions*	*Assets Billions*	*Net Income Billions*
1. General Motors	$24.3	$14.9	$1.7
2. Standard Oil (N.J.)	$14.9	$17.5	$1.0
3. Ford Motor	$14.8	$ 0.9	$0.5
4. General Electric	$ 8.0	$ 6.0	$0.3

SOURCE: "The 500 Largest Industrial Corporations," *Fortune* (May 1970).

What is involved here is not simply an escalation in quantity. With automobiles the largest single commodity item in the United States, General Motors operates plants in seventy American cities in 18 states. There are seven plants in Canada as well as plants in twenty-three other nations. The magnitude of the General Motors bureaucracy rivals that of any foreign government, with thirty-one major operating divisions. General Motors accounts in whole or part for the payroll of over 7 million Americans with a direct employment of 794,000. The General Motors Acceptance Division, extends nearly as much short-term commercial paper in the United States as the U.S. Treasury itself. The influence GM has upon the prosperity of all parts of the economy is not bound simply to market mechanisms. The corporate board members of GM exercise their influence over virtually every major corporation in America, holding directorships in such a variety of companies as AT&T, U.S. Steel, IBM, Mobil Oil, and Gulf Oil. Their influence dominates both industrial and financial institutions, including the First National City Bank of New York, the Morgan Guaranty Trust Company, the Mellon National Bank and Trust Company, the First National Bank of Boston, the First National Bank of St. Louis, the

National Bank of Detroit, the Royal Bank of Canada, the Montreal Trust Company, and numerous others.

The implications of this economic empire say much about the structure of the American economy. General Motors averages as much profit per car as it pays out in wages. For every dollar of increased labor cost between 1947 and the 1970 strike, it has raised automobile prices by $3.75. The sustained profit "target" is 20 percent on capital after taxes. This is achieved operating thirty-six of the fifty-two weeks of the year.

> If General Motors had been satisfied in 1965 with a 13 percent rate of return on investment — the after-tax profit rate enjoyed by the average manufacturing corporation — it could have raised the wages and salaries of every one of its employees everywhere in the world by nearly $3,000. Or it could have cut the wholesale price of every car and truck by about $300, permitting a substantially greater cut in retail price.[4]

Public policy is one product of the automotive industry. Another is culture. While it is today popular to cite the immense financial magnitude of General Motors, few stop to take notice of the more subtle, but every bit as potent, influence GM has had upon our everyday lives. Originally sought as a means of traveling between two points, the auto has absorbed both means and ends, and today it is used as much for eating, watching movies, banking, making love, courting, establishing status, mailing letters, listening to the radio, watching TV, "getting away from it all," sleeping, camping, hunting, living, and, not least, dying. C. A. Doxiadis estimates that an American male between twenty and fifty-nine spends one-third of his total free time commuting. What is involved here is not simply a deteriorating quality of life, but the basic alienation of life itself, such that little basic experience today takes place without the automobile.

Moreover, the automobile is a key instrument in maintaining the capitalist ethic of individualism. The extent to which the automobile encourages and feeds upon individual competition is recognized by anyone driving in a heavy traffic jam, nudging, pushing, inching his way ahead of the next person. In describing the nature of democratic despotism, de Tocqueville recognized that "despotism, which by its nature is suspicious, sees in the separation among men the surest guarantee of its continuance, and it usually makes every effort to keep them separate."

The separation of people caused by the use of motor vehicles takes an enormous toll on the quality of social life. Not only does it diminish human contact with other members of the society, but also within families. The hours spent commuting for many persons is a major distraction from family life, separating parent from child, men from women. On a larger level, the highway-freeway system is a classic tool in the maintenance of class and racial barriers between communities. More than one ghetto in America has found itself circumscribed by highways. As such, highways have become a major mechanism of class division, separating black from white, rich from poor. We will consider other implications of the highway system later.

iii

If the form of American life is increasingly dependent upon the automobile, much of the substance of commodity production today results directly or indirectly from petroleum feedstocks.

In the nineteenth century the development of coal brought with it fundamental transformations in the fabric of society. In the twentieth century petroleum determines the fabric of production and consumption more than any other industrial influence.

The average person has almost no contact with petroleum. It is something transferred from a pump to his car, or sometimes one will hear of a bad oil spill. However, the gasoline produced for the auto is a fraction of the use to which petroleum is put. A single gallon of petroleum yields a vast variety of products. Of that gallon, 43.3 percent goes for gasoline, 37.4 percent for fuel oil, 4.3 percent for kerosene, 2.3 percent for jet fuel, 1.3 percent for lubricants, 1.3 percent for petrochemicals, and 10.1 percent for all other products.

While the percentage of crude utilized from a single barrel for purposes other than gasoline is relatively small, the value in terms of commodity products produced is overwhelming in scope and variety. From petroleum three thousand different products are produced with another three thousand products produced from petrochemicals. Petrochemicals result from the fact that crude petroleum is a mixture of thousands of different hydrocarbons — or compounds of hydrogen and carbon. The proportion of hydrogen to carbon compounds varies in accord with the particular geology of the petroleum location.

The petroleum chemist alters the composition of atoms in crude petroleum through a variety of processes which yield an almost infinite number of new products from the original hydrocarbon base. An even partial listing of products provides a marked respect for the importance of petroleum to the fabric of our present culture; fuel oils, lubricants and greases, solvents, synthetic fibers, films, synthetic rubber, plastics, pharmaceuticals, paints, detergents, perfumes, explosives, herbicides, aspirin, dyes, fertilizers, ink, metal alloys, cement — to name just a few. The same people who bring us DDT and other dangerous pesticides also bring the nondisposable plastic bags and the constant pollution of the air we breath.

The complexity of the petroleum market cannot be understood without considering the remarkable development of the petroleum industry itself. For it is only with the extension of petroleum capital that the extension of petroleum commodities is made possible. Capital formation, in this sense, underlies the basic ecological responsibility of the petroleum industry. Capital concentration affects both social and biological stability.

Petroleum was accumulated in the geological structure of the earth over a period of 500 million years. While oil was recognized bubbling up out of the ground thousands of years ago, it was not until 1859 in Pennsylvania that the first commercial oil well was dug. During the 1870's John D. Rockefeller established a complete monopoly of the American oil industry. The hegemony the Rockefeller family exercises in the world of capital, petroleum exercises in the energy domain. By the turn of the twentieth century, Standard Oil controlled three-quarters of the U.S. refining capacity, virtually all of the oil pipeline and over 80 percent of all petroleum sales. At this point the original trust was reorganized by John D. as the Standard Oil Company (a New Jersey corporation) and it began extensive overseas investment. In a landmark antitrust suit in 1911, the Supreme Court upheld a lower court decision which dissolved Standard Oil of New Jersey into thirty-four legally separate enterprises. In 1927 Jersey Standard divested itself of directly owned and managed affiliates and became essentially a holding company.

Today the Rockefeller network encompasses the various Standard Oil companies: Jersey, Indiana, Ohio, California, and Mobil (formerly Standard Oil of New York), and other institutions such as the Chase Manhattan Bank, Eastern Airlines, Metropolitan Life Insurance, Equitable Life Insurance, for starters. What ties these companies together is the controlling interest of the Rockefeller family. The family owns together a controlling 15 percent interest in the various Standard Oil companies (Jersey 13.5 percent; Mobil 16.3 percent; California 11.9 percent; Indiana 11.4 percent), as well as 10 percent of the stock of the Chase Manhattan Bank (the major source of petroleum capital). One estimate of the Rockefeller family wealth is put at $15 billion, with another $90 billion in controlling assets.

The 1911 antitrust decision against Jersey Standard constituted little more than a minor setback. As John Moody described the situation in the *Masters of Capital,* "Such dissolutions have proved in the end, however, to be mere changes in form, for the various companies involved continued to be owned, controlled, and managed by practically the same men, with little if any real competition." Today this empire operates under hundreds of different names through thousands of different organizations in most every country of the world.

In 1969 the domestic net sales of the petroleum industry (35 companies which process 97 percent of all crude oil in the United States) amounted to $58 billion with net income at $5.8 billion.[5] Aside from the requirements of motor vehicles, aviation and other forms of transport, petroleum is refined in the United States by 275 refineries for space heating as well as the basic feedstock for thousands of petrochemical products. This petroleum is marketed in over 30,000 terminals and 225,000 retail gasoline stations across the United States. The petroleum industry employs one out of every 60 people in the United States. In 1967 the *New York Times* reported that the 33 petroleum companies among the 500 largest manufacturing and service companies accumulated over one-third the total earnings of all five hundred. The First National City Bank of New York, in a survey of 2,250 manufacturing companies, noted that the 96 oil producers and refiners constituted one-fourth the income of the entire 2,250.

iv

There is no better example of what in an earlier chapter was described as political capitalism than the corporate influence of oil and automobiles upon American public policy, both domestic and foreign. Through a complex variety of programs and regulations the Federal government has all but become the primary guarantee of the power and position of the automotive and petroleum industries.

More here is simply involved than pressure, or market mechanisms. The intimate ties that have developed between large corporations and government are without question more significant than those that exist between big and small business. This is precisely the consequence of the merger between corporate and state capital.

In recent years two presidents of automobile companies have become secretaries of defense; Charles Wilson of General Motors and Robert McNamara of Ford. Wilson brought with him Roger Keyes, his vice-president, to be deputy secretary. Keyes later returned to GM as a vice-president and then director. Thomas Mann, ranking member of the State Department, became president of the Automobile Manufacturers Association. George Romney, a president of American Motors, became secretary of Housing and Urban Development. Former secretary of commerce, John T. Connor, became a director of General Motors. These are only a beginning.

There is no better example of the magnitude of this influence than the privileges extended to the automotive industry through the Highway Trust Fund. Highway expenditures are the classic example of the manner in which Federal spending serves to consolidate the hold of private business upon the public sector. Between 1944 and 1961 the Federal government allocated its entire transportation budget to roads and highways. Today, nearly 20 percent of all government spending other than defense is for highways and related projects. The Highway Trust Fund was originated in 1956 to finance the interstate highway system. It is scheduled to go out of existence on September 30, 1972, but it is unlikely to do so since the fund is virtually self-perpetuating. The fund comes from a Federal tax of four cents a gallon on gasoline as well as a lesser excise tax. This money can only be utilized to build new highways. Since the Federal government pays 90 percent of the costs of construction, the states are pressured to forever expand their highway construction program in order to acquire the Federal funds and associated employment and revenue. Federal highway revenue today is about $6 billion, expected to increase to $6.5 billion over the next several years.

The National Organization of State Highway Officials has claimed that highway needs for the next fifteen years will run $320 billion — a figure which compares to the current national debt of $375 billion! President Nixon has asked Congress to extend the fund for another four years. In 1969, the Federal government spent approximately $50 on highways for every dollar spent on mass transit. To date the interstate highway system has cost the American taxpayer $32 billion, while Federal expenditures for the collapsing railroads approximate less than $1 billion. This, of course, helps explain why there are today nearly 15,000 highway and street construction contractors.

The expansion of the highway system is merely one more example in which government serves to consolidate, then increase the market of a particular industry. The dependence upon cars establishes the dependence upon oil as well. The more roads that are built, the more dependent the society becomes upon roads, automobiles, oil, and the vast culture they entail: suburbs, take-out food, motels, repair shops, parking violations revenue, pollution control industries, road maintenance . . . the list is virtually endless. The government itself receives no small share. The special state and federal taxes dependent upon the present transport "system" are telling:

Tax	*1969*
Special State and Federal Motor Tax	$16.0 Billion
Special State and Federal Truck Tax	$ 5.1 Billion
Highway User Tax Revenues	$ 9.4 Billion
State Gasoline Taxes	$ 5.9 Billion
Federal Excise Tax Paid by Motorists	$ 6.7 Billion

SOURCE: *1970 Automobile Facts and Figures,* pp. 60-65.

The dimensions of this income are impressive. Special motor user taxes contribute 20 percent of all state tax revenues. Total government receipts for highways in 1970 constituted well over $20 billion, or one-tenth the national budget. That means that one-tenth the national income can be spent for nothing but asphalt jungles.

The highway system is a critical cog in the centralization of economic and political power. The vast system of patronage it supports is the most extensive lobby in the Capital. The prohibitive provisions of the Trust Fund prevent other forms of transportation from being developed. When citizens in California attempted to pass a resolution in 1970 to divert gasoline taxes from highway construction into mass transit, considerable sums of money were spent by the highway-automobile-petroleum companies to defeat Proposition 18. The illegal campaign contributions by these companies that defeated Proposition 18 are now under investigation by the state attorney general. This pattern has been repeated in several other states.

Highways result in 14 million crashes annually at a cost, in addition to human life, of $25 billion in damages. Yet much of this wanton death and destruction could be eliminated in a matter of months. Today many safety devices exist which could all but eliminate the 55,000 highway deaths a year as well as the hundreds of thousands of injuries from the auto. Technology exists today which could transform the automobile into a relatively safe and nonlethal form of transportation. Numerous inventions exist for both the interior and exterior of the vehicle, including antiskid braking, air bags, interior cocoons, crashworthy frames and innovative bumpers and noncontact dash boards. More sophisticated forms of computerized vehicle travel and operation along automated highways are already available.

The Department of Transportation reported in November 1970 that the nation's auto population is growing twice as fast as its human population as compared to three times the population growth in the 1950's. Here, perhaps, is the population control demanded by the conservationists. While 6,000 people are born daily in the United States, about 12,000 automobiles are produced. With a growth rate of 4 percent, there should soon be as many automobiles in America as people. Within this context we can understand the concern about air pollution and emission standards. Says Russell Train, chairman of the Environmental Quality Council,

> The data on current levels of pollution in our urban environment . . . make it clear that even the application of more stringent standards in 1975 will only prolong the downward curve in vehicle emissions until the middle 1980's. After that, carbon monoxide levels will again rise because of the sheer number of automobiles on roads and highways.

The first auto pollution control devices were developed by engineers outside the auto industry. When "strict" California standards required Detroit to lower emission standards, they discovered overnight a "clean air package." We noted earlier in Chapter Two the consent decree of October 28, 1969, which the big four auto manufacturers signed with the Justice Department regarding a conspiracy to stall in the installation of pollution control devices. And yet, a pollution free engine has been available for decades.

V

Numerous forms of steam propulsion have been around for years — remember the Stanley Steamer? Calvin and Charles Williams of Ambler, Pennsylvania, built a steam-powered auto nine years ago.

Other forms of Rankine cycle engines, as they are called, have been around for forty years. Such an engine would practically eliminate present day emission problems. Consider the following:

Contaminant	*Internal Combustion*	*Williams Steamer*
Hydrocarbons	900 parts per million	20 ppm
Nitrogen oxides	1,500 ppm	40 ppm
Carbon monoxide	3.5%	0.5%

Moreover, the steamer emits no lead and requires no clutch, no transmission, no carburetor, which would almost eliminate repair costs.

A variety of other propulsion systems have been devised. David W. Rabenhorst, supervisor of special projects for Johns Hopkins University's applied physics laboratory has developed a "flywheel energy storage system" which is light, silent, cheap, and produces no exhaust at all. The wind-up principle of the power system has been known for over five thousand years, though the present version is more sophisticated than previous attempts. The power unit would operate by a free-spinning disc, or flywheel, which would store the energy required to drive the car. The heavy flywheel would be respun by a minimal electrical current applied when the car is not in use. This system would cost one-tenth of one cent a mile to operate, as opposed to the two cents a mile in gasoline costs of the present automobile. Top speed would be seventy miles an hour. Detroit yawned at the idea and thought Rabenhorst crazy, obviously preferring the present internal combustion engine.

The Clean Car Race of August 1970 from Massachusetts to California demonstrated the technological simplicity of developing a nonpolluting engine given the will. Four Wayne State students equipped a 1971 Ford with pollution control devices which beat the 1980 Federal control standards. Said Wisconsin Senator Gaylord Nelson, "It is nothing short of hilarious that the largest concentration of engineers and experts in the world can't match the wits and creativity of a group of college students." In another case, Wallace L. Minto of Florida sought to sell rights to his Freon-powered Rankine cycle engine to American auto firms but found no interested parties. Minto then went to Datsun. The Japanese intend to produce the Freon engine in 1972 with mass production beginning in 1973.

The immense social costs of the automobile in terms of the health and general welfare of the public are staggering to contemplate. California physicians have found that mortality rates from arteriosclerosis, heart disease, and cerebrovascular disease to be 79 percent higher in polluted areas than in those with relatively nonpolluted air. Hydrocarbons are considered largely responsible for the astonishing rise in lung cancer in urban areas over the last half century. Mounting evidence suggests that the asbestos from brake linings also presents a particularly hazardous threat to the human lungs.

A report by Lester B. Lave and Eugene P. Seskin in *Science*, August 20, 1970, provides chilling indications of what happens every day to the internal chemistry of your body:

> Approximately 25 percent of mortality from lung cancer can be saved by a 50 percent reduction in air pollution.... It seems likely that 25 percent of all morbidity and mortality due to respiratory disease could be saved by a 50 percent abatement in air pollution levels.... There is evidence that over 20 percent of the cardiovascular morbidity and about 20 percent of cardiovascular mortality could be saved if air pollution were reduced by 50 percent.... Finally, there is a good deal of evidence connecting all mortality from cancer to air pollution.

We are here talking about mere *reductions* of the present air pollution concentrations. All evidence suggests that the conditions of health and disease within urban society are fundamentally influenced by atmospheric pollution. Reduction in the contaminants, while a start, is not the objective. They must be eliminated. The constant congressional objective of controlling, or regulating the emissions misses the point altogether.

Considered by some to be one of the greatest inventions of American science and technology, in reality the automobile has been responsible for more death and human misery than any other single factor in American life. The automobile is the single largest cause of death for people between the ages of fifteen and thirty-four. It is a major killer among all other ages. It should be obvious from the above that to attempt to force Detroit to install a clean engine or regulate emission standards is bound to fail. Any significant transformation of the automobile represents nothing less than the reconstruction of the present American economy — of which automobiles are the center. Only a fundamental reconsideration of the purposes and use of the automobile in terms of the total issue of transportation and structure of the society will curb the present biological and social costs of the internal combustion engine.

The basic issue of transportation today is that beginning in 1940 an excellent diversified system of transport was usurped by the automobile and airplane. Up until 1940, the United States had spawned a flexible and diversified transport system; railroads and airlines expanded what was originally a canal and river system. In addition, from the late nineteenth century on, a variety of transport systems were devised both within and between cities using electric trolleys, steamers, ferryboats, motor trucks, motor buses, and individual motor vehicles. During the thirties the railroads made numerous technical innovations including electrification, the changeover to diesel power, and other improvements which greatly enhanced regional and continental transportation. Local service was neglected, due in large measure to World War II and the concern about transporting war supplies.

The exclusive use of the automobile for intercity transit and the airplane for long distances has all but crippled what was at one point a highly diversified system. The increasing dominance of the military since World War II in all forms of research and development favored the highly technical and long-distance orientation to air transit. Within recent years, the aerospace industry has been virtually controlled by the military requirements for travel, the B52 spawning the 707 and the C5 the 747. Today, instead of adequately exploring numerous short-range possibilities of air transit, we spend vast sums on the SST. Similarly, the ascendency of the petroleum and automotive industries as the major political and economic force, together with their most affluent offspring, the Highway Trust Fund, have prohibited the development of diversified transit even when technology existed or could have been readily developed. A diversified transport system will not be developed unless the forces that prohibit it, such as the military, the Big Four, and Boeing, are fundamentally altered or eliminated.

Present-day government regulation only serves to further the monopoly of auto and airplane and protect the interests of Detroit and Seattle. The current financial troubles of the railroads and airplanes results from the economic conditions of inflation, in part, and the illusory notion of "common carriers." This notion essentially provides that Federal regulation insures that carriers must serve both profitable and nonprofitable markets. The theory is that high-profit operations subsidize low-profit operations. Given the economic crunch over the last few years, high-profit merchandise is carried at the expense of low-profit goods and people. Thus the major airlines eliminate small communities from service. The failure of cross-subsidization is rooted in the structure of transport motivation. Profit first, service second, or not at all. As a result, many private firms and small communities must generate their own form of transportation, or each individual must arrange his own transportation. Of course this fosters diversity, but at the sake of balance. At the heart of the matter is the virtual monopoly that the auto and long-distance air traffic has upon all requirements for transportation. Put another way, as long as the economic monopoly is permitted to exist, so will the cultural monotony.

The solution does not lie simply in the introduction of more mass transit proposals. In the past twenty-five years the number of people carried on the nation's mass transit systems, according to the American Transit Association, has declined by two-thirds, from 19 billion to 6 billion! It is precisely because transportation treats people as a faceless mass, that patronage has declined. The inoperable levels of congestion which modern urban centers exhibit present a fundamental problem of scale and fragmentation. This congestion and fragmentation results essentially from the present-day patterns of capital control. Namely, because communities are deprived of their own resources, in the form of skills and taxes, individuals are forced to seek income outside the community in which they live. Carried further, the concentration of capital in the inner city core, in terms of business and industry, further consolidates the centrality of the civic center and the colonial status of the suburbs.

The answer here must not be to improve the transit of people from outside to inside the city and back again like clockwork. The issue at stake is one of generating a balance of capital and resource distribution, such that local communities can provide work and income for their members, and not require them to commute to and from other political jurisdictions. Put another way, decentralization must occur. Decentralization both for reasons of scale, simply because Manhattan and other urban centers are virtually inoperable, and decentralization because this is the best and most reasonable way to achieve a diversified form of transportation.

What this entails is an integrated transportation system designed to maximize public welfare, not private profits. As such the purposes of movement itself will be called into question. The shipments of commodities will be diminished because local neighborhood and transneighborhood political jurisdiction could achieve increasing autonomy. What is desired is neither the single person mode of transport nor mass transit which moves a faceless mass from suburb to city, back, forth, back, forth . . .

The question of a decent and workable transport system cannot be separated from the issues of capital and resource distribution and concentration. The accentuated separation of work and living space that now exists is the inevitable consequence of the capitalist mode of accumulation. This will not be overcome merely by more subways. The diversification of transport and the decentralization of population will require political autonomy, autonomy achieved by local political organization and struggle. To believe for a moment that national superplans will solve complex local issues would be foolhardy.

It is obvious that an industrial society can cope with transit problems if there is a will. Take the high-speed Japanese train — an experiment vastly superior in conception and design to the Metroliner between Washington and New York. The difference there was will, not technical superiority. The Japanese simply determined to spend the required $2 billion. While the problems are of a different magnitude here, they are not insoluble, except if we imagine that the present apparatus, which got us into the plight, can get us out.

vi

The complete dependence of American transportation upon the internal combustion engine when innumerable other forms of transport are available is a feat of Detroit comparable to the petroleum companies' generating a monopoly for the vast bulk of the energy budget — when many other sources of energy are readily available.

If there is a single force responsible for preventing the development of a diversified, balanced, and ecologically sound system of energy, it is the petroleum companies. Senator Lee Metcalf has called this the "galloping oligopoly in the energy business." The penetration of the largest twenty-five petroleum companies into the fuel industry is marked by the shift in perception that has occurred in most oil boardrooms. Not satisfied with controlling 75 percent of the present energy budget, they have been moving over the last five years to become "energy companies." What this means is the elimination of effective competition between oil, gas, coal, electricity, and uranium.

Standard Oil of New Jersey since 1965 has become one of the two largest holders of coal reserves and a principal operator in uranium — the source of nuclear power. Eighteen of the largest twenty-five oil companies have invested heavily in the production of uranium. Though oil now accounts for roughly one-sixth the production of uranium, they hold 45 percent of all known uranium reserves and constitute more than half the annual new discoveries.

Natural gas is geologically linked to the discovery of petroleum. All major petroleum companies have natural gas operations, which have become substantial little sisters to oil in the energy market. The two largest owners of coal, Humble Oil (a New Jersey Standard subsidiary) and Continental Coal (a Continental Oil subsidiary) are petroleum companies, which account for 25 percent of the nation's coal. The National Economic Research Association has compiled data for the Senate antitrust subcommittee regarding the energy monopoly:

> Not only are such firms buying competitive fuels, but because of their size they are in a strong position to outbid other kinds of industries that wish to venture into the energy field. The acquisitions of the oil companies, across the energy market spectrum . . . may be viewed as a classic horizontal integration on a scale comparable to the formation of trusts in the latter decades of the nineteenth century. In short, the oil companies, themselves portraying their activities as efforts at diversification, are in fact systematically acquiring their competition.

It is within this context that the cry of "power crisis" must be understood. For if anybody has pulled the plug out, it is the energy monopoly itself. We noted in Chapter Two how the electric utilities are in large measure responsible for much of the bulk increase in electrical consumption. In 1968 and 1969 the previous annual increase of 5 percent spurted to 9 percent. This in large measure can be attributed to the increased advertising by utilities for appliances, and the like, not a rise in the population.

A critical effect of the energy monopoly is the ability to determine energy costs. Petroleum has always set the price. This is achieved through a variety of means. The import quota protects domestic supplies from cheap foreign gluts. At home, oil companies in cooperation with oil-producing states set "allowables" or the amount of production in accord with the "nominations" of the oil companies. Theoretically this works as a means of conserving natural resources. In fact, it works to regulate the local production of petroleum in accord with the monthly requirements of the companies. As if this were not enough, the companies then determine the price of crude by "posting" what they will pay for oil in each of the five major producing regions of the United States. Do you know of any other market in which the buyer sets the price?

Shortages are easily manufactured. The scarcity of natural gas in 1971 might be understood by the simple desire on the part of the petroleum companies for price increases. By not hooking up 500 potential wells in the Gulf of Mexico to distribution pipelines, power becomes scarce.

While oil is used in only 9 percent of U.S. power plants, coal is used in 53 percent. As noted earlier, coal is increasingly the domain of the large petroleum companies. In the early sixties coal producers cut back on their production, supposedly because utilities were turning toward the construction of nuclear power plants. Consequently the utilities failed to obtain adequate long-term coal contracts. At present, nuclear power supplies 4 percent of the energy budget. Today over a dozen nuclear power plants are stopped in midconstruction by law suits brought by citizens concerned about the hazards of atomic power. Another four dozen plants will no doubt receive such suits when they are far enough along in the construction process. The shortage of coal, together with the obstacles to the nuclear plants, rising technical difficulties resulting from poor management of the Northeast power grid and the promotion of increasing Reddy Kilowatt by the utilities themselves, brings us to the current "power shortage."

Coal production was about 5 million tons short of demand in 1970. With exceptions, slack in production of coal in the United States coincides with the acquisition of coal companies by the petroleum giants. In any events, given the present constitution of the energy monopoly, oil cannot help but win. Public utilities have placed all their chips on nuclear power. If they begin operation, oil will be selling them virtually all of the uranium. While demand for refined uranium in 1969 was 5,000 tons, it is projected to leap to 35,000 tons in 1980. In effect, the monopoly of oil then will dwarf the dimensions of Standard Oil in 1911.

The oil companies blame the power shortage on the utilities and the instability of the Middle East. The utilities blame the conservationists. The conservationists blame the consumer. The source of the manufactured power crisis in fact is the nature of energy production itself: the monopoly of ownership and the monotony of the oil-centered energy budget. Only by attacking the structure of political and economic power at the root of this dilemma will solutions be found.

A rush to nuclear power, given the current controversy about the hazardous implications, can only further complicate any possible solutions. The promise of nonpolluting energy from nuclear power is elusive, at best open to deep consideration. Says David Freeman, White House energy advisor, "The fossil fuel plant is a lot dirtier than the nuclear plant. At the same time it may be that fossil fuel plants are an immediate threat to the environment while nuclear plants represent the long-term threat for the future."

The obvious question arises, is it possible to generate a safe and nonpolluting form of energy? If so, why hasn't it been developed? If we cannot expect the energy companies to undertake such research, what about the institutions of higher learning?

Senator Lee Metcalf noted in the December 28, 1970, *Congressional Record* that fifty-three universities control almost a billion dollars worth of voting stock in energy corporations. Metcalf cited as a consequence the difficulty Federal officials had in obtaining University of California consultants in the wake of the Santa Barbara oil spill of 1969, "Some university experts did not want to endanger their consulting arrangements (with companies) and university grants." The holdings of the fifty-three universities look like this:

	Number of Shares	*Value as of 12/5/1970*
Oil companies (44)	11,487,949	$526,773,290.00
Electric utilities (85)	10,963,272	321,590,645.00
Gas utilities (32)	1,805,683	57,204,063.00
TOTAL (161)	24,256,904	$905,567,998.00

SOURCE: News release from the office of Senator Lee Metcalf, 28 December, 1970.

Standard Oil of New Jersey attracted the largest investment, worth $129 million by such universities as Brown, California, Columbia, Dartmouth, Lehigh, MIT, Northwestern, Pittsburgh, Princeton, Syracuse, Texas, Vanderbilt, Case Western Reserve, Williams, and Yale. Next were Texaco, Gulf, Mobil, and Standard of California.

This may help explain why no university in the United States has a course of study about the political economy of the petroleum industry — the largest corporate body in the world. Moreover, there is not a single book published on the vast empire of Standard Oil.

Not a single university in the United States has undertaken a serious program for the development and testing of nonpolluting forms of energy. Nor have they undertaken any long-term serious research about alternative forms of transportation. And yet, knowledge exists and has been recognized for some time that the way out of the current energy squeeze is available, if we wished to pursue it. No simple solutions about curtailing consumption will be adequate, though that is important. What must be addressed are the forces which are at the root of our present patterns of consumption, not the mere quantitative figures. Our dependence upon fossil fuel, for an example, is not bound to the geological formation of the earth but the social formation of capital. An authentically diversified system of energy would challenge not only the hegemony of the oil companies, but the very structure of society. Let us consider the possibilities.

Fossil fuels, whether they be petroleum, natural gas, or uranium, are nonrenewable sources. In other words, they are of a limited supply, and are known as capital sources. In contrast, there are the income sources of energy, sources which are continually being replenished. These sources include:

> *Photosynthesis*: There are other ways than by its food energy role in which energy may be directly extracted from vegetation, e.g., through fuels from wood or by microbialaction in biological fuel cells.
> *Direct Solar Energy*: The sun's rays may be used in a variety of ways, e.g., to power boilers and cooling devices by means of lenses and reflectors, or by converting them directly to electricity by photoelectric cells or photochemical fuel cells.
> *Hydrological Energy*: The flow of rivers, the rise and fall of tides, the waves of the sea, and the hydrological cycle of evaporation and precipitation are all replenishable sources of energy that can be exploited by suitable means. The ultimate source of such forms of energy is a combination of indirect solar energy and gravitation.
> *Wind*: Though this is intermittent and variable, improvements in energy storage methods may enable this source to be more widely used.

Temperature: Temperature differentials between the atmosphere and the earth (or the temperature and the sea) yield energy potentials of considerable magnitude.
Geothermal Energy: Volcanic sources of hot gases and waters may be exploited either through natural vents or by drilling artificial vents.[5]

We tend to consider these possibilities as remote and impractical. In fact, in America, we know almost nothing about them. Is this due to the inapplicability of such sources of energy or the current monopoly on the market?

The point to be stressed is not that we will magically find one truly clean source of energy that will replace the current strangle on fossil fuels. But rather that a wide diversity of sources do exist and are currently being experimented with, and that such sources could be adapted, based upon geographical potentials, to provide a safe and almost nonpollutant supply of energy. Such a development would depend upon fundamentally different political relationships. To provide some indication of the public ignorance, we can consider geothermal energy as one possibility.

The U.S. Geological Survey estimates that as much as 1.3 million acres of land, mostly within the western part of the United States, may be available for geothermal development, with an energy source of from fifteen to thirty thousand megawatts. This is an area which connects the states of Washington, Oregon, California, Montana, Idaho, Nevada, Wyoming, Utah, Colorado, Arizona, and New Mexico. This is a fraction of the earth's capability, and may well be a conservative estimate within the United States. At present, the United States produces 83,000 kilowatts annually from natural heat; only Italy and New Zealand produce more. Federal estimates suggest that this could be expanded ten times under current economic conditions.

Geothermal energy was not utilized to produce electricity until early in the twentieth century. However hot waters were employed by the Romans and was put to commercial use first in Lardorello, Italy, in 1777, to produce borax and other chemicals for industrial purposes. In the United States, Boise, Idaho, utilized steam issuing up from the earth to heat homes and greenhouses in 1890. In the

1930's Klamath Falls, Oregon, employed steam for space heating. Today, a fraction of the Pacific Gas and Electric output is produced from geothermal sources. In Iceland, hot-water energy was employed commercially in 1925 for space heating, and New Zealand, recognizing their importance, nationalized geothermal sources in 1946. Today, numerous other nations, most notably Japan and the Soviet Union, have given development of geothermal energy high priority in their national planning. The Japanese, if you remember, are also developing the nonpolluting Freon engine that Detroit abandoned.

John Lear, science editor of *Saturday Review,* states that "geothermal power is attractive because it can be generated much more cheaply than power from any other source; furthermore, under proper management, it is capable of enhancing rather than deteriorating the environment.

Beginning in 1962, Senator Allan Bible of Nevada initiated a measure to encourage explorations of geothermal energy. Eight years later it passed Congress. In October 1970 a UN Conference on geothermal energy was held in Pisa, Italy. It had initially been proposed for Riverside, California. Washington was not interested. As Lear put it, "oil and steam don't mix." The heat required for steam dissolves the hydrocarbon compounds of petroleum. However, this is but the geologic picture. Politically, geothermal energy is but one of many alternative sources of energy that could be developed and produced in the United States, energy that would challenge the position and power of the petroleum empire. It is expected that Richard Nixon will veto the Bible bill that would open U.S. lands to geothermal development. Lyndon Johnson, no enemy of oil, vetoed a similar bill in 1965. The authoritative *Electrical World Week.* of December 21, 1970, stated that "the battle to open government lands, which contain the bulk of America's geothermal reserves (some 500 times the heat potential of gross U.S. coal reserves) has been hard fought by utilities, oil companies, and independent geothermal developers." A "grandfather clause" of the bill would permit developers who had previously invested substantially in public lands, to convert oil and gas leases by simply matching the highest bid that the Department of the Interior receives from other developers. Win or lose, oil wins.

Our failure to develop geothermal energy is but one illustration of the domination by petroleum companies over the fabric and workings of the society. Energy and transportation are two critical components, but they are by no means the entire story. The fact, for instance, that 70 percent of the people live on less than 2 percent of the land is directly involved in the considerations of geothermal energy. For discoveries of a cheap, clean, and directly abundant source of energy would reduce the dependence of the population upon the oil and automobile companies for energy, transportation, and a variety of other factors which force people to live in the highly congested urban cores. In this manner, the concentration of capital within the petroleum and automobile industries is responsible for the concentrations of people, and thus the increasing monotonization of all aspects of everyday life, of which transportation, energy, and land-use patterns are but examples.

vii

The organization of society for the ends of the petroleum and automotive industries is sustained by a complicated and diverse network of Federal support and assistance. The Highway Trust Fund was considered earlier, which serves to bolster both autos and oil. The petroleum industry, however, is unmatched in the variety and complexity of special Federal privileges. The extensive network of government incentive to petroleum is rooted in the development of the Federal regulatory and tax structure throughout the twentieth century. While particular figures have changed, such as the reduction in the depletion allowance from 26.5 to 22 percent, the principle is the same. C. Wright Mills noted of the petroleum privilege, "The important point of privilege has less to do with the percentage allowed than with the continuation of the device long after the property is fully depreciated." While conservationists and oil critics strive to muster sufficient steam to budge half a dozen key tax privileges, any number of alterations in the books of a major producer can offset any particular decrease in quantitative terms by the overriding magnitude of privilege that is involved.

There are basically three kinds of incentives given to petroleum by the government, prorations, import controls, and tax advantages. They practically amount to the same thing, but are merely applied in different situations and diverse formulations. The most popular is the depletion allowance, originated in the income tax law of 1913, and raised to the classic 26.5 percent figure in 1926. The theory of the depletion allowance, established first by petroleum and then extended in various forms to other natural resources, is that a percentage of profitable production should be written off against taxation. The reasons are, the oil companies argued, (1) that oil is a critical natural resource to the welfare of the society and thus production must be encouraged, and (2) that because the oil is depleted when produced, it cannot be considered actual income, but capital, because the oil is destroyed (consumed) in the act of profit taking. In any other business this would be considered the simple cost of doing business. Oil thought otherwise. The purpose of the depletion allowance, wrote *Business Week* in 1961, is to "permit . . . tax free recovery of the value of deposits." As a consequence, the depletion allowance works to reduce taxes, and thus provide a substantially higher margin of profit. This profit is taken not once, but every year upon an investment which was made just once, the exploration and drilling of a well. This means that the depletion allowance is not taken in regard to individual wells, but, due to special tax write-offs, annually against the collective costs of doing business. As long as production continues, somewhere on the order of 50 percent of the net profit is thus tax free. Treasury Department studies indicate that oil and gas depletion deductions are ten times what the actual cost of depletion was to have been.

During the 1960's alone the depletion allowances cost the U.S. taxpayer on the order of 20 billion of pretax oil income. This amounts to $1.3 billion a year, as indicated by a Treasury Department tax reform study in 1968. The 26.5 percent depletion allowance, originated in 1926, may have cost the U.S. Treasury $140 billion in lost income — income which the petroleum companies can utilize for any number of projects ranging from the TAPS Alaskan pipeline to investment in the Middle East.

There has been a long, consistent roster of opposition and support of the depletion allowance in Congress. While the allowance was reduced from 26.5 to 22 percent, this hardly matters in the profit margins of the petroleum companies, who simply shift around their figures under some other tax provisions they have obtained.

The import quota system is another monumental petroleum incentive. It was inaugurated in 1959 by President Eisenhower (born a Texan) under the justification of national defense. The origins of the quota in fact resulted from extensive pressure brought to bear by domestic producers who were being pushed out of domestic markets by giants such as Jersey Standard and Shell with gluts of cheap imported oil. Put simply, the cost of domestic crude is artificially maintained by a limitation on the amount of oil that can be imported, since Middle East oil is both more productive and less expensive than domestic sources. Under the mandatory import law, the government issues "import tickets" to U.S. refiners worth about $1.25 a barrel, the difference between U.S. and Middle East crude at the East Coast. These tickets are essentially currency which the Department of Interior issued to the domestic refiners to establish a parity between domestic and import oil.

The quota system costs the American public somewhere on the order of five to seven billion dollars a year in inflated costs. Concretely that means that a family of four may pay for fuel oil an excess of one hundred dollars annually in New York, and perhaps two hundred and fifty dollars in Wyoming.

These are only some illustrative examples of petroleum privilege. The Senate Small Business Committee concluded in 1949 about the proration system, "A perfect pattern of monopolistic control over oil production and the distribution thereof . . ." There exists today few such perfect examples of collusion between big business and government to control prices and production as the proration system. The present foreign tax credit all but allows capital investment abroad to escape taxation.

As shown earlier, large sectors of American society are concerned with the preservation of the present system of privilege and incentive to the petroleum companies. Under such circumstances it is not surprising that Richard Nixon on February 20, 1970, ignored a suggestion by a former presidential task force to abolish the import quota system. Nixon promptly dismissed its chairman, former Labor Secretary Schultz, and replaced him with Attorney

General John Mitchell, a former law partner of Nixon's serving petroleum interests. Referring to the depletion allowance in Texas, Richard Nixon vowed, "As President, I will maintain it." Robert G. Sherrill in *The Accidental President*, documents some of the lessons about "black power" from the career of Lyndon Baines Johnson.

This complex array of tax arrangements is quite lucrative for the petroleum companies, and the source of much of their power. The effective tax rate for all manufacturing is more than twice that of petroleum. In the five-year period between 1963 and 1967, the five major American companies' U.S. income tax rate was 4.9 percent; taxes paid to foreign governments were over 20 percent of their net profits before income tax of $20 billion. In 1968, average corporate income tax was 40 percent upon earnings, for petroleum it was less than 8 percent. In 1964-1967 with earned profits in excess of one half billion, Atlantic Richfield effectively paid *no* Federal income tax, ARCO said about its fabulous North Slope find that "the tax incentives to exploration had accomplished what they were intended to do."

The necessity of this planetary scramble on the part of the petroleum companies for increasing profits and supplies was put in perspective by the November 10, 1969, issue of *Oil and Gas Journal*, the industry's own trade journal. In the history of the industry some 225 billion barrels of oil have been produced, less than half the estimated present reserve. In the United States and abroad the petroleum industry has known reserves of 203 billion barrels and expects to discover another 473 billion. In addition, improved production is estimated to yield another 300 billion barrels of oil, with a growing recovery factor of 40 percent up from the present 30 percent. This would mean that an estimated 775 billion barrels of oil is probably now available without ever drilling another wildcat well — or enough for fifty-four years given the present growth of consumption. Even at expected increased rates of production, the world will have ample supplies of oil until 1990 without any further exploration. These estimates do not include the Southeast Asian deposits and are likely to be short in estimating Soviet and Canadian fields.

The momentum for the ravenous development of petroleum today is rooted in the extensive system of special privileges the government grants to the petroleum industry. The momentum behind the exploration and development of oil determines the present policy in Alaska and Machiasport, the destruction of the atmosphere, housing, and land, and the present military presence in Indochina. In order to bring together several of the vast implications, we can consider one petroleum company as a microcosm. For this purpose we will consider one of the minors, which reveals in a telescopic way the facets and characteristics for which petroleum is today known world wide.

viii

There can be no comparison in sheer economic, political, and cultural power when considering either Standard Oil of New Jersey or General Motors. But it is perhaps by comparison with a smaller, less-known operation, that one begins to understand exactly the magnitude of the colossus.

Atlantic-Richfield (ARCO) is the eighth largest oil company in the United States and tenth in the world. It ranks as the twenty-ninth largest industrial in the United States with combined sales and assets for 1969 amounting to $6.9 billion with $227 million in profits, sixteenth among U.S. industrials for net income. ARCO originated in 1877 as the Atlantic Oil Storage Company, part of the gigantic Standard Oil empire. Standard officially relinquished control in 1911 upon Federal dissolution. In 1966 the name was changed to the Atlantic Refining Company and in January 1966 Richfield Oil was merged into the company. Then in March 1969 Atlantic-Richfield merged with Sinclair Oil Corporation which doubled the size of the original operation and quadrupled sales. This clearly put ARCO into a strong competitive position with the five majors. Besides the exploration, development, and marketing of petroleum products, ARCO is engaged in petrochemicals, atomic fuels, fertilizer distribution, and plant food manufacturing. With five U.S. refineries, ninety-six gasoline plants, and 24,000 retail outlets, ARCO recently spent $60 million at the cost of $3,000 per station to face-lift the old Atlantic or Sinclair sign to read ARCO. Part of the new face entails construction of the tallest building in the cities of Philadelphia, Dallas, and Los Angeles.

Besides domestic operations, ARCO has foreign marketing operations in South America and Europe with primary properties located in the United States, on the North Slope of Alaska, Venezuela, Iran, the United Kingdom, Libya, and Indonesia. All told, ARCO involves some ninety-three total or partially owned subsidiaries.

ARCO is a company on the make. The chairman of the board, Robert O. Anderson, has declared: "I'm interested in seeing ARCO become the number one oil company in the world, not necessarily in size but in performance." This hunger for growth, control, and consolidation is not unique to ARCO. It is the natural tendency of the energy monopoly to want to acquire increasing control over the resource base it depends on. This explains Standard Oil of New Jersey's expansion in the early part of this century in regard to natural resource deposits in the Middle East and the current competition among oil giants for the vast oil deposits on the Alaskan North Slope and Indochina.

ARCO pioneered the now legendary oil discovery at Prudhoe Bay on the North Slope of Alaska and has a 27.5 percent interest in the Trans Alaska Pipeline proposal, as well as a substantial interest in the Manhattan oil tanker venture. ARCO to date has ordered three of the largest American flag oil tankers (940,000 barrels) under construction. Besides this, ARCO was the first petroleum company to announce it would market a no-lead gasoline in 1970. To round out the picture, ARCO was the ninety-second largest defense contractor in 1969 and received the second largest contract for aviation fuel in the industry.

While there can be no doubt that the dynamics of what makes ARCO grow are to found first in the logic of the industry per se, one cannot disregard the personality and importance of the people behind the individual corporations, more important with a minor such as ARCO than of course with the five majors. As much as any force, Robert O. Anderson has been instrumental in directing the phenomenal growth and development of ARCO. One begins to understand something of the dimensions of petroleum power with a short sketch of Anderson himself. The son of a petroleum magnate in Chicago, Anderson has been chairman of the board of ARCO for five years and was the guiding force in the "new look" to ARCO, the merger with Richfield and Sinclair. In addition to his chores at ARCO Anderson is the largest individual landowner in the United States, his territory is larger than the family-owned King Ranch of Texas. He is also a National Republican committeeman from New Mexico, a trustee of the University of Chicago and the California Institute of Technology, and a board member of the Columbia Broadcasting Company and the Chase Manhattan Bank. Together with the Security First National Bank of New York, Chase is the major financial institution behind the empire of oil. Anderson is deeply involved in national political campaigns. Anderson himself contributed $44,000 to the 1968 Republican campaign. Some twelve of the thirty-three ARCO executives contributed to that campaign in the amount of $66,000, $65,000 to Republicans and $1000 to Democrats. As noted earlier, ARCO is involved in the development of atomic fuels. Of the eighteen Atomic Energy Commission contractors who contributed to political campaigns in 1968 who were not among the top Department of Defense contractors, ARCO leads the pack. Perhaps in recognition of Anderson's performance, he received a 44 percent rise in salary between 1968 and 1969, from $156,250 to $225,000.

The direct influence of such a man and his corporation should not be discounted. The law firm of Richard Nixon, prior to his presidency, handled the affairs of Robert Anderson and ARCO. The connection between these matters and such decisions as the Alaska oil development and the contracts of the AEC together with the ARCO nuclear fuel interest, should not be discounted.

Anderson is also heavily involved in cattle, real estate, mining, and other investments. He is a big-league philanthropist and on the board of directors of the Aspen Institute in Aspen, Colorado. He was chairman of the board of Resources for the Future and a chief financial banker for David Brower's John Muir Institute and the Friends of the Earth, the former named after the founder of the Sierra Club.

This is but a brief yet illustrative sketch which begins to pinpoint the parameters of power that one of the minor petroleum companies in the United States can play. While other companies may be less flamboyant and less personal, their sphere of influence is no less extensive, no less important.

ix

This influence can determine the geological composition of the Alaskan tundra or the pattern of all American investment abroad. More so than automobiles, petroleum operates as a single world economy, with the overriding share of American foreign investment. One third of all U.S. foreign investment is petroleum, 71 percent of American investment in the Third World and 25 percent in Western Europe. Something of the magnitude of what is involved can be grasped by considering Standard Oil of New Jersey. This multinational corporation has an annual operating budget of $15 billion, employing 150,000 people around the world in some 250 subsidiaries. They control an 18.3 percent interest in U.S. oil in the Middle East and 17 percent of all U.S. sales in Europe. Jersey Standard operates thousands of affiliates and joint ventures in a hundred nations. Such popular brand names as Humble, Enjay, and Esso are but a few of the numerous Jersey Standard marketing brand names. If the annual budget of Jersey, Indiana, Ohio, and California Standard were considered as a nation, they would rank third behind the domestic economy of only the United States and the Soviet Union.

Of all forms of American foreign investment, petroleum operates in a unique context. While other forms of development invariably offer secondary and tertiary kinds of employment and growth in the country involved, petroleum by comparison does not. Because of the highly capital intensive nature of the industry, petroleum tends to generate high capital enclaves and elite labor pools amidst a sea of squalor and poverty. The oil capitals of the world, Tehran, Venezuela, Jakarta, for example, illustrate the dichotomy between a prosperous urban center and a surrounding rural belt of intense poverty. This may have no small consequence in the recent trends toward nationalization on the part of oil-producing countries. While all forms of economic colonization reveal unequal distributions of wealth, oil in particular seems to severely extenuate divisions within the society it enters between wealth and poverty, overdevelopment and underdevelopment. Michael Tanzer in *The Political Economy of International Oil and the Underdeveloped Countries* summarizes his definitive study noting that "the economics of the international oil industry generally brings the profit-maximizing goals of the companies into direct conflict with the economic development *cum* political independence goals of the underdeveloped countries."

The uneven development that petroleum brings to developing nations entails both political and ecological implications. The extenuated enclave form of urbanization generates massive disparities between urban and rural, with modernization occurring around the capital enclaves of petroleum income without adequate social and cultural guidelines to insure the preservation of cultural autonomy. While transportation in developing nations is often poor, it is common to see excellent roads connecting oil fields and ports, with surrounding squalor. Centralization of industry and other basic necessities generates ecological imbalances both in terms of population congestion, as well as the other ecological results of intensive industrialization.

Petroleum is more at the root of development around the world than any other single factor. This process relates directly to comments in Chapter Two about the rise of political capitalism. For, if there is any force within the business community that is at the root of American public policy abroad and home, it is the petroleum and automotive industries. The extent of this interplay is beyond normal imagination.

The composition of government at home and abroad is determined in no little measure by petroleum and automobiles. While the list of oil Congressmen is extensive and well publicized on occasion, the influence of petroleum contributions to local elections, and the influence they exercise against certain candidates is unparalleled.

In the 1968 campaign, of the five top corporate contributors, two were auto and one petroleum, GM, Ford, and Atlantic Richfield. The two largest personal contributors were Henry Salvatori, in oil as well as Litton Industries (a major space contractor), and Richard King Mellon of General Motors and Gulf. As we have seen with Charles Wilson and Robert McNamara, more than one key cabinet official has come from the ranks of automobiles or petroleum. Robert Anderson, the chairman of the board of Atlantic-Richfield, was the last man seen to leave President-elect Richard Nixon's suite before Nixon announced the appointment of Walter Hickel as Secretary of Interior. Anderson's Atlantic-Richfield and Jersey Standard are the pioneers in Alaskan oil development.

The structures of domestic and international policy are both dependent upon the extensive influence of the Rockefeller oil empire around the world. Four of the last five secretaries of state were Rockefeller employees. Dean Acheson, Truman's secrctary of statc, was president of the Rockefeller Foundation and also a member of the Standard Oil law firm of Sullivan and Cromwell. John Foster Dulles was secretary of state for Eisenhower and a senior partner in Sullivan and Cromwell. At the same time, his brother, Allan Dulles, senior partner in Sullivan and Cromwell, was director of the Central Intelligence Agency. During that period the United States was instrumental in the overthrow of governments in Iran and Guatemala where Standard Oil of New Jersey was deeply enmeshed in domestic political disputes. John Foster Dulles, who died in 1959, was replaced by Christian A. Herter, who had married into a Standard Oil family. Dean Rusk, secretary of state under both Johnson and Kennedy, served as president of the Rockefeller Foundation before he became secretary of state. During this period the United States invaded Cuba, where Standard Oil had the dominant American interest. The record — four of the last five secretaries of state. Henry Kissinger, President Nixon's primary advisor for National Security Affairs, was a longtime policy advisor to Nelson Rockefeller and director of the Rockefeller Special Studies Report. The Rockefeller Council on Foreign Relations has supplied more high-level foreign service personnel than any other constituency. The pattern of petroleum influence upon government policy is far-reaching and unrivaled.

The extent to which the petroleum companies and the U.S. government work hand in hand was well demonstrated in the recent controversy between the ten-member Organization of Petroleum Exporting Countries and the large major oil companies. Richard Nixon, on a day's notice, dispatched the number-two man in the State Department, John Irwin, to pressure the government of the Persian Gulf States. Irwin was a former partner in a Rockefeller associated law firm, a director of the Rockefeller controlled U.S.

Trust Company of New York and former legal counsel for the Rockefeller Foundation. In 1967 the nationally syndicated columnist Jack Anderson described the relationship, "... The State Department has often taken its policies right out of the executive suites of the oil companies. When Big Oil can't get what it wants in foreign countries, the State Department tries to get it for them. In many countries, the American embassies function virtually as a branch office for the oil combines." In large measure, this results from the fact that the world petroleum market is virtually an American cartel. Five of the seven majors are American: Standard Oil of New Jersey, Royal Dutch Shell, British Petroleum, Texaco, Gulf, Mobil, and Standard Oil of California. These five American giants control 70 percent of the non-Communist supply of petroleum in the world, down from 90 percent in 1950.

Oil and natural gas constitute three-quarters of the industrial world's supply of energy. With a quarter of the world's population, industrial nations consume 81 percent of the world's petroleum and 95 percent of the natural gas. The United States, with 6 percent of the world's population, consumes 34 percent of the world's oil and gas. Aside from oil and gas, the industrial nations consume 77 percent of the world's coal production and 80 percent of the hydroelectricity. The Far East and Asia, with one-half of the world's people, consume less than one-tenth of the world's petroleum, yet that is where the bulk of the world's oil is located. That is also the arena for much of the world's troubles. Petroleum today is at the center of both of the world's major combat zones, the Middle East and Indochina.

An indication of what is involved is provided by an examination of the 1970-71 conflicts between oil-consuming nations and the ten-member Organization of Petroleum Exporting Countries (OPEC). The place to start, of course, is with the role of the United States. The United States is both the largest national producer of petroleum and the largest consumer. Annually the U.S. consumption increases by 3.6 million barrels per day, or the daily equivalent of any large Middle East producing nation. Of the 14.8 million barrels per day that the United States consumed in 1970, 20 percent was imported. About 6 percent comes from Canada, 3.5 percent from Venezuela and about 3 percent from the Middle East. In other words, the

United States is not dependent upon the Middle East stores for domestic needs.

While the U.S. energy consumption climbs at 6-8 percent, the six Common Market nations soared by 13 percent last year (France by 17 percent) and Japan by 13 percent. Europe imports 75 percent of the energy demand and Japan 90 percent — almost all from the Middle East. Middle East oil up until just recently has been unilaterally controlled with a 59 percent interest on the part of American petroleum companies (Esso 18.3 percent, Gulf 9.5 percent, Texaco 8.7 percent, California Standard 8.1 percent, Mobil 5.3 percent, et cetera). At the same time, OPEC oil constitutes the bulk of U.S. companies' foreign production (Esso 91.9 percent, Gulf 86.9 percent, California Standard 93.9 percent). As a consequence, Middle East oil is critical to American petroleum interests to maintain their structure of profit from the sale of OPEC oil to Europe and Japan. Standard Oil of New Jersey made 50 percent of its 1970 profit from foreign investment — mostly in the Middle East. The American military presence in the Middle East has not refrained from taking those measures necessary to protect U.S. oil interests, including continued support for the British where required, a continual supply of arms to both sides of the Middle East conflict, and the outright use of force when necessary — whether it be through CIA assistance to military coups or the landing of troops in Syria and Lebanon to protect pipelines from Saudi Arabia and Iraq to the Mediterranean. CENTO, the Central Treaty Organization, was a diplomatic device promoted by the United States in the Middle East to keep the Soviet Union out of the area — though most of the Middle East nations found more danger from the United States than the Soviet Union, and refrained from joining.

The momentum for OPEC resistance came largely from concern about a pro-Israeli policy on the part of the United States. The leading U.S. antagonist, Libya, clearly spoke of the desire to "hurt" the U.S. companies until they prevailed upon the government to stop the "Zionist aggression." One U.S. response was for Attorney General John Mitchell to suspend the appropriate antitrust laws so that the petroleum companies could operate as a single team. However, there is no specified legal authority which gives the attorney general the right to suspend such laws. Only the Supreme Court could initiate such a ruling.

The domination of the Middle East economy by U.S. petroleum interests has been a steady concern of the Soviet Union. They are rapidly increasing their assistance to Arab nations in order to encourage the development of national petroleum companies that could one day achieve independence from American companies. In this context, one might venture a guess that the decision in 1970 to install Sam 3 missiles in Egypt at the same time as the brewing OPEC controversy may not be unrelated.

The growing concern on the part of the Western oil empire was expressed by Walter J. Levy, recognized as the "dean of the oil consultants," in December 1970 at the OPEC meeting in Caracas, "It is most foreboding in terms of the Western economic and strategic interests. There is a definite potential that producing nations will take unilateral control over the pricing and taxation of all foreign production. If so, the consuming nations are in for a rude awakening." Indications are that the February 15, 1971, Persian Gulf settlements will cost the consuming nations dearly. In the United Kingdom, following the February 15 settlement, across the board barrel costs have risen 42¢, on top of an 87¢ a barrel hike in 1969. That may amount to an additional $144 million a year to the UK's import bill for 1971. In Germany, gasoline products costs were raised about 24¢ a barrel, and that was just the beginning. Similar hard times are ahead for Japan, Italy, Spain, and India. In the latter case, India even before the February 15 settlements was running an import bill of 180 million annually for oil, or 9 percent of the total import bill. The rise may go to $216 million for petroleum products. The closing of the Suez Canal together with the new OPEC tax and price demands cost the petroleum companies an estimated $3-5 billion a year, which they no doubt will pass on to the consumers.

Given the voracious pattern of petroleum consumption projected for the future, new stable sources of petroleum have been a constant concern since World War II. Alaska and to some extent Nigeria have proven hopeful to date, but without question the largest potential are the vast petroleum reserves of Southeast Asia and the Western Pacific.[6]

On December 1, 1970, the Thieu-Ky Government of South Vietnam issued Public Law No. 011/70 regarding the "Exploration and Exploitation of Petroleum and Related Taxation, Fees and Exchange Regulations." This law signaled the official go-ahead for negotiations on offshore leasing arrangements between South Vietnam and twenty-one contending petroleum companies, mostly American. Extensive American petroleum exploration in Indochina and South Vietnam has been conducted for over two years under the auspices of the United Nations Economic Commission for Asia and the Far East (ECAFE).

The eighteen offshore South Vietnamese concessions are located in the Gulf of Thailand and the southeast offshore region adjacent to the penal colony of Con Son. The government will offer the concessions on a "best offer basis," allowing great flexibility on the part of the contending petroleum companies. The specified tax rate of 50 percent can be lowered by official decree to 45 percent or raised to a maximum 55 percent, though not retroactively. In addition to the offshore area, prospecting has also occurred in the rich Mekong River Delta and on the alluvial plains of Thailand. On February 24, 1971, Thailand's Senate passed two petroleum laws which cleared the way for initial offshore test drilling for six companies, five of which are American. Potential petroleum deposits have been detected along a continental shelf which connects Japan, Indochina, Indonesia, and Australia. With over one million square miles of jungles and shallow seas, this is the largest continental shelf in the world. Also, it appears, the largest deposit of petroleum.

American interest in the natural resources of Southeast Asia is long standing. Richard Nixon and Dwight David Eisenhower referred regularly in the early fifties to the "valuable deposits of tin and prodigious supplies of rubber and rice." In 1965, Henry Cabot Lodge, U.S. Ambassador to South Vietnam, stated, "He who holds or has the influence in Vietnam can affect the future of the Philippines and Formosa to the East, Thailand and Burma with their huge rice supplies to the West, and Malaysia and Indonesia with their rubber, oil, tin to the South. Vietnam thus does not exist in a geographical vacuum — from it large storehouses of wealth and population can be influenced and undermined."

The escalation of American economic investment in all of Indochina was underscored on November 11, 1969, in a statement made by the U.S. Ambassador to Thailand before congressional hearings: "There is one very large development that may take place. Nobody knows at this point whether it will prove out or not. That is the oil exploration going on in the Gulf of Thailand. . . . Now if that develops, it of course will be a very major additional investment in the country." Subsequent to the coup of General Lon Nol and the U.S. invasion of Cambodia, Khmer-Thai economic negotiations have increased substantially. These negotiations center on a common plan for petroleum exploration.

Aside from the role in the determination of overall war policy in Southeast Asia, oil may as well play a strategic position in determining military objectives. Testimony given by the United States Air Force before the Electronics Battlefield Subcommittee of the Senate Armed Forces Services Committee suggests that a major reason for the recent invasion of Laos by South Vietnamese and American military personnel is the destruction of a petroleum products pipeline running out of North Vietnam just north of the DMZ in Southern Laos. The existence of the pipeline was disclosed in Senate testimony before the committee on November 18, 1970, by Brigadier General William John Evans. This pipeline appears to have played a major role in the North Vietnamese troop movements along the Ho Chi Minh Trail, supplying an estimated 3,000 six-wheel heavy Russian trucks with fuel. The Air Force had previously disclosed in Saigon that within the last two-and-one-half years the portion of the trail open to trucks in the dry season has been extended from 350 to 1,550 miles.

It appears that the military escalations in Cambodia and Laos coincided with the actual determination of petroleum deposits in Indochina. A press release of the French wire service Agence France Presse of February 7, 1971, states,

> The very recent discovery of important oil deposits in Thailand and South Vietnam, explains in great part, according to reliable sources in Moscow, the resurgence of military activities in Indochina, particularly the recent events in Laos. . . . One cannot know the precise moment the

existence of the deposits became known to the Americans, but according to the same sources, one cannot discard the possibility that decisive information on this subject was obtained before the unrolling of events in Cambodia, last spring. [Author's translation.]

Jacques Decornoy, Southeast Asian editor for *Le Monde,* wondered in the January 8, 1971, edition, "Have the oil companies perhaps received some solid assurances from Washington concerning the United States' willingness to "hold" Indochina, and South Vietnam in particular?" [Author's translation.]

Already American petroleum investment in Southeast Asia is extensive. Standard Oil and Shell have had a long-standing investment in the region for over fifty years. At present offshore oil concessions dot the map of Southeast Asian waters. By the end of 1971 Indonesian petroleum production is expected to reach one million barrels per day, with American interests dominating 80 percent of production. U.S. investment in Indonesian oil has sprinted from $100 million in 1969 to $130 million in 1970 and is expected to climb to $160 million in 1971. Pertamina (the national Indonesian petroleum company) has negotiated more than twenty-eight production and exploration contracts with foreign companies, seventeen of which are American.

Over $100 billion worth of military spending has been poured into South Vietnam, with another $4 billion in nonmilitary aid into Saigon in the last twenty years. The extensive network of highways, harbors, railroads, airports, and communications provide an ample foundation for an oil rush to Southeast Asia. David Rockefeller of the Chase Manhattan Bank estimated in May 1970 that American oil firms would spend $35 billion on petroleum in the Western Pacific over the next dozen years. To date there have been five major petroleum finds on the shelf, one by Royal Dutch Shell and four by American companies: Atlantic-Richfield (who pioneered the Alaskan North Slope discovery), Cities Service, the Union Oil Company, and Natomas of California. Aside from Indonesian waters, there has been a rush to the Gulf of Thailand and Malaysia with continued attention to the coastal waters of South Vietnam.

The *Wall Street Journal* of September 22, 1970, reported rumors that Standard Oil of New Jersey (Esso) had discovered oil in its twenty-eight-thousand-square-mile concession off the coast of Malaysia, directly adjacent to the South Vietnamese blocks. To date Esso has made no announcement, realizing that news of the find would greatly boost the price of South Vietnamese bidding. However, Esso opened a greatly enlarged petroleum refinery in Singapore on February 19, 1971. A U.S. Embassy official in Singapore, base for the burgeoning petroleum operations of Southeast Asia, suggested, "We've had all the feelings of an oil boom here in Singapore already. But with the recent chaos in the Middle East the oil moguls must be frantic now to get more firmly into Southeast Asia." These events will no doubt shape the destiny of the entire Pacific Basin.

The American balance of power in the Pacific relies upon Japanese dependence on American petroleum supplies. While the Japanese have declared their intention that by 1985 at least 30 percent of oil imports should be brought in by Japanese owned resources, major American companies own substantial interest in Japanese firms. Negotiations between the Soviet Union and Japan for oil and gas are on the upswing. The Soviet oil and gas discoveries in the Siberian Arctic are no doubt an important spur to American exploration and development in Southeast Asia, given its proximity to Japan. Soviet supplies will more than likely dwarf Alaskan finds, and given current North Slope delays, easily compete for the Japanese petroleum market.

x

The structure of the automotive and petroleum industries is more than simply an important part of the American grain. The hundreds of billions of dollars which the major automobile and petroleum companies control are virtually beyond the influence of or benefit to the American public.

It would be a mistake however to separate big business from government, to assume that either business runs government or that government dictates the priorities of business. Something else entirely is involved. States one executive vice-president of Standard Oil of New Jersey,

> We have a very good knowledge transfer mechanism. We pick up things in Japan, Venezuela, and Australia and bring them here. And a bigger company can afford more good people. We get them from all countrics and all kinds of backgrounds. Somehow, I feel, *the whole is greater than the sum of its parts*. (Emphasis added.)

While the maneuverings and roles of the various parts themselves are of fantastic concern, it is about the whole that the ecology of capitalism is concerned: not simply better government, more responsible business, better emission standards, or a lowering in the depletion allowance. The nature of corporate capital is somewhat like the body of an amoeba: if a part is cut off, the flow of protoplasm is directed to form another appendage. As long as it is attacked piecemeal, it will continue to flourish and refurbish the lost parts with new extensions. The whole, in this case, is more than the sum of its parts.

VI

The Conditions of Liberation

The political pestilence of the United States is today undoing the work of organic evolution. The political and economic inequities of capitalism are today directly responsible for the biological traumas of the earth's natural systems and its population. Seen in this light, the accumulation of carbon dioxide in the earth's atmosphere or mercury in the oceans is tantamount to the accumulation of capital by the American corporate structure.

An illustration of the consequences of this form of social organization is the fact that within the United States 75 percent of the people live on less than 2 percent of the land. That 2 percent of the land is a cancer ward.

There can be little doubt but that "medical care is largely unrelated to the health status of the population."[1] Within the urban context it is clear that poverty and air pollution have more to do with the health and welfare of the population than anything the medical profession espouses. It must become equally obvious that the conditions of health and disease are today the ultimate political issues, issues which can find resolve only in political and economic reconstruction of the urban complex.

The spectacle of modern urbanization results directly from the extreme irrationality of centralization and consolidation, in which people, like capital, are concentrated on a massive scale in order to efficiently monopolize production for private profit. The disequilibrium of capital concentration is analogous to the disequilibrium within the individual personality from crowding, as demonstrated by J. B. Calhoun in his experiments with rats. New York is a paradigm of this insanity: the squalid atmosphere, the constant biological invasion from stress of every imaginable kind, the racial separations into ghettos, the breakdown of every major social service system, and the alienating forms of inhuman conduct are not the exception but the rule. The fact that many New Yorkers (like most inhabitants of America's large cities) are unable to accept this today is a symptom of the terminal stages of urban degeneration. We are unable to accept the reality of death because we have never known the reality of life.

The homogenization and banality of all aspects of life are the motor forces of present-day urbanization. Marx understood this when he noted that "in a certain sense, all of history can be thought of as a struggle between the country and the city." The systematic domination of man over woman and man over nature is rooted in the six-thousand-year-old legacy of scarcity, patriarchy, social hierarchy, and the mechanistic world view. These social attitudes do not achieve their ultimate expression until the beginning of the seventeenth century, when the excesses of the capitalist era exacerbate the ultimate contradiction between human and biological survival.

The present imbalance within natural systems — the human destruction of the biosphere — originates in the imbalance within human society. The possessive, authoritarian, and mechanistic behavior of the capitalist society is translated directly into human contact with the environment. Human and biological hierarchy are inseparable. A society in which one is exhibited invariably exhibits the other, they are two sides of the same coin.

Consequently, proposals which seek merely to alter or transform the methods or means of hierarchical structure, which seek to merely substitute one set of rules or rulers for another, cannot possibly abort the present course of destruction. There must be no illusions about the consequences of pursuing the present course of legislative and judicial action against the chemical and biological warfare of monopoly capitalism against all living forms.

ii

It has been a primary purpose of this book to illustrate the overriding responsibility of capitalist society in the global structure of social and ecological disorder. Britain, West Germany, and Japan are simply junior partners. The extent of American responsibility is illustrated in the realization that Los Angeles

County alone has more registered automobiles than the entire continent of Africa, or that the state of California consumes as much electricity as the 750 million people of China.

While it is commonly countered that socialist countries exhibit their own social and ecological disorder, this skirts the basic issue at stake: is the capitalist organization of society inherently anti-ecological? Can the social organization of natural resources — in terms of profit, private property, and competition — achieve any reasonable harmony with the life support systems of the earth?

The evidence I have gathered suggests not. Moreover it must be realized that America as an empire is today the primary and most substantial agent of biological and social destruction. While it is true that ecological imbalance occurs within the Soviet Union or the People's Republic of China, there is one fundamental difference: the basic *systemic difference* between the capitalist and socialist view of man and society.

With a productive output of between two-thirds and three-fourths of that of the United States, a comparable population, and a land mass two and a half times that of the U.S., the Soviet Union has demonstrated an impressive record of ecological balance. The most serious ecological threat to the Soviet Union is water contamination. Not to be understated, industrial wastes have severely damaged many rivers and inland waterways. Forestry and the effects of hydroelectric station construction are current political issues in the Soviet Union as well. However, the staid *Newsweek* was moved recently to observe that "the Soviet Union is a healthy distance ahead of most industrialized countries in environmental control."

Socialist society considers ecological and social balance as inseparable. Consequently, the Soviet Union maintains a declining birth rate and a relative balance between rural and urban settlement. Air pollution is virtually nonexistent in the Soviet Union. This is not simply because there are less industry and more land than in the United States, but results first from the strict Soviet planning policy of separating population settlement from industrial production sites, and secondly, from the fact that there are only some 5 million automobiles in all of the Soviet Union. Public transportation in the Soviet Union is the rule, not the exception. Five major Russian cities have renowned subway systems, with another ten systems under construction. Urban planning in the Soviet Union provides for a federally financed, electrically powered subway system for every city with a population exceeding one million.

Though Moscow has only 80,000 privately owned automobiles, thousands of these have already been converted from gasoline to liquefied natural gas. The entire heating system of Moscow, like that of London, has been converted from soft coal to natural gas. While the Soviet Union boasts the second largest petroleum industry in the world, after the United States, it projects an energy budget within a decade which relies on geothermal power for somewhere on the order of half its energy.

The importance of ecological balance to the planning policy of the Soviet Union is not a recent fad. Lenin himself recognized the importance of ecological considerations in the development of the Soviet Union, formulating a draft program of research for the Academy of Sciences. Within socialist society, ecological balance is but one aspect of the overall concern for social balance.

While it is true today that the socialist nations have emphasized the importance of economic output and modernization, unlike capitalism, socialist planning might just as well be utilized to curb ecologically improper forms of growth and development. Perhaps the most complex of examples, and the most important, is China. In a highly provocative paper, "Capitalist and Maoist Economic Development,"[2] Professor John Gurley, Department of Economics at Stanford University, illuminates the basic distinctions between capitalist and Maoist economics.

There is considerable infantile worship and illusion about China amongst the Left today, but this must not blind us to the deeper lessons of Chinese Communism. We have already considered at some length the implications of capitalist development. To recapitulate, the market economy is incapable of addressing the origins of ecological imbalance. For such issues affect all of us, not merely the buyers or sellers. Our relation to capital must not determine our absolute dependence upon the life support systems. An economic system motivated by profit and competition cannot address the common welfare. The logic of industrial development under capitalism serves only the needs dictated by capital.

People within this context are merely a means to an end. The end is the accumulation of capital, not the maximum development of human beings. People are considered one input into production, along with capital and land.

Because capital accumulation is the measure of all things, capital accumulation also determines the basis of societal organization. Capital determines privileges and hierarchical relations. People are commodities. Similarly, the air, land, and water become the captives of capital accumulation. The aggrandizement of capital is the equivalent of the ownership of the earth. Consequently, as Marx noted, "the production of too many useful things results in too many useless people." Today the health and welfare of the American people — and the people of the world — declines proportionally to the soaring GNP.

Capitalism can be characterized by the classic uneven swings between boom and bust, the excessive production of superfluous commodities to the deprivation of health and education, the continued extenuation of inequities between wealth and privilege, and the increasing homogenization and specialization of all experience. Progress is achieved through individual entrepreneurs who privately own the means of production. Success is measured by the national output, or GNP. As portrayed first by Adam Smith and developed later by Keynes, output is the paramount end of development. Human welfare is equated with economic acquisition and output, and is the basis of Madisonian interest politics. While competition is declining with the massive intervention of the state, "capitalist development, even when most successful, is always a trickle-down development."[3]

Gurley identifies the basic components of Maoist economic development as "central planning, public ownership of industries, and agricultural cooperatives and communes." Because China is presently striving to increase production rather than limit output, questions about investment versus consumption, trade, the allocation of material inputs, and prices are basically determined by the state. However, within this context a great deal of decentralization and diversification occurs. This is because unlike capitalist development, material output is but a means to an end, not the end itself.

The primary concern of China today, while mirrored in economic development, is the development of human beings in their full, creative potential. The creation of the "universal" or "socialist" man is the primary concern and motivation for economic development. Maoist thought is here descended directly from the Marxist preoccupation with the evils of specialization, bureaucracy, and alienation.

Mao, more than anyone in recent times, has promoted the belief that "All genuine knowledge originates in direct experience.... There can be no knowledge apart from practice." Consequently, the development of Communist men and women can only be accomplished through struggle and participation, through the collective practice of the Chinese people. As a result, the welfare of each is inseparable from the welfare of the many. If development is to occur, it must involve everyone on an egalitarian basis. Gurley notes that "Economic development is not likely to occur unless everyone rises together. Development as a trickle-down process is therefore rejected...and so they reject any strong emphasis on profit motives and efficiency criteria that lead to lopsided growth. Their emphasis, in short, is on man rather than 'things.'"

The composition of the proletarian world view is inherited basically from the Marxist-Leninist theory of dialectical materialism. In *On Contradiction,* Mao explains that "The development of things should be seen as their internal and necessary self-movement, while each thing in its movement is interrelated with and interacts on the things around it."

The emphasis upon flux, struggle, selflessness, nonspecialization, constant participation, and balance have profound ecological and social implications for our present concerns. Inherent within dialectical thought is the concern for balance and relativity. The social balance which is a motivating concern of China today is analogous to the natural balance achieved within an ecosystem where indeed the welfare of the individual is absolutely dependent upon the welfare of the many, as opposed to the capitalist view which promotes the welfare of the individual at the expense of the many.

One illustration of the implications of this view is in the Chinese principle of "multiple use," which is the guiding factor of Chinese industrial development. The notion of "multiple use" manifests the Chinese concern for decentralization, diversification, and the nonspecialization of human and material development. The development of an industrially "integrated complex" seeks to ensure that "one man is specialized in one skill and familiar with others," and that "one machine does many jobs" in order to raise productivity while curtailing investments and costs. This results from the underlying philosophy of dialectics which drives industrial development in Communist China. One example is the view of industrial "waste." The Chinese view waste and what is valuable as the "unity of opposites in a thing." In other words, "In given conditions, 'waste' can be transformed into what is valuable and the useless into the useful." The unity of opposites and the overriding concern for the integrity of the whole which characterizes dialectical thinking is also the foundation of ecological science. This extended quote from *Peking Review* should illustrate the operation of Chinese industry in regard to waste:

> ...Nature's material resources cannot be fully utilized by producing one product. In making one product, resources are partially transformed into this product and the rest becomes "waste." The question is how to look at this "waste" — from which point of view and with what kind of attitude. From the metaphysical point of view, waste cannot be used and should be got rid of. On the contrary, the materialist dialectical view holds that what is waste and what is not waste are relative terms. There is nothing in the world that is absolute waste. "Waste" under one condition may be valuable under different ones. "Waste material" left from one product can become a good material for another product. After being transformed and utilized, "waste material" can become a product or useful material.[4]

This dialectical attitude is responsible for greatly increasing Chinese production in recent years, such that at the same time that new construction and expansion of industrial development occurs, a maximum effort is maintained to fully develop and diversify existing industry. By contrast, McGraw-Hill Publications'

Department of Economics estimated in 1970 that it would cost $144.5 billion currently to replace "technologically outmoded facilities."

The Chinese conservation of resources entails a primary concern for the maximum utilization of existing forms, rather than the rapid proliferation of commodities and equipment merely in order to maximize profit. The attitude involved is of critical importance to development:

> All things gradually become old or damaged in the course of use. This is the natural law of the development of things. After an all-around analysis of damaged equipment, workers have become aware of the fact that something that is damaged is bad in one respect but this does not mean everything about it is bad. After being repaired or restored, old or damaged equipment becomes greatly changed and very serviceable.[5]

Inherently, Chinese industrial development eliminates the massive irrationality of annual style change and planned obsolescence. The conservation of resources, both material and human, is integrally intertwined. The maximum use of material resources is predicated upon the modest, frugal living of Chinese culture. As a consequence, concern for ecological balance flows directly from social balance and diversification. An ecologically sound scale and level of production are determined in accord with maximizing public welfare, and thus the greatest enrichment of all participants and each area involved in the whole.

Balance is clearly the motivating principle of Chinese development. One example is the industrial city of Shenyang in Northeast China. A few years ago, seventeen or eighteen provinces and cities supplied Shenyang with grains and vegetables. Socialist construction seeks a balance between city and countryside, industry and agriculture, workers and peasants. Consequently, Shenyang set out to become self-reliant in agriculture, rather than extending the specialization of industry. Within the framework of a unified arrangement for the entire city, cultivation was undertaken in a systematic way. The consequence was increased production, not inefficiency, "While achieving self-sufficiency in grain and vegetable supply," the city increased the area under production by three times from 1966 to 1970. The nature of the balance is thus expressed:

> Industry and agriculture are interdependent and promote each other in their development. Industrial development can promote agricultural mechanization and modernization and push forward farm production. Agricultural development will provide industry with grain, raw materials, and markets and lay a firm foundation for the development of the entire national economy.... Through such mutual support and promotion, industry and agriculture will make common advances.... Mutual support of industry and farming and mutual promotion of the city and the countryside constitute the economic base of the worker-peasant alliance under the socialist system.[6]

In this manner, the concentrations and specialization characteristic of capitalism do not occur. And, contrary to popular assumption, productivity is increased by diversification.

The massive anti-Communism of the last two decades, the distorted reporting of the Cultural Revolution and the narrow empiricism of most Western observers has all but blinded Western scholars and the general public from an honest view of China. It is normally pictured as a land of seven hundred million hungry, ill-clothed, and ill-housed people when nothing could be further from the truth. Over the last twenty years all basic necessities, most notably health and education, have improved steadily and compare today with standards in all Western nations. Gurley notes that "it would not be farfetched to claim that there is less malnutrition due to maldistribution of food in China over the past twenty years than there has been in the United States."

China cannot be understood except within the framework of the Marxist-Leninist goal of Communist man in a classless society. This is a view of the world so fundamentally different from ours as to require the most disciplined care in attempting to understand Chinese society.

Gurley's evaluation of the past twenty years concludes that "Communist China is certainly not a paradise, but it is now engaged in perhaps the most interesting economic and social experiment ever attempted, in which tremendous efforts are being made to achieve an egalitarian development, an industrial development without dehumanization, one that involves everyone and affects everyone."

The ecological implications of this form of social development are staggering to contemplate. What in essence has been demonstrated is that a highly productive society with an industrial rate of production since 1950 of 11 percent, including one-quarter of the people of the world, has begun in fact to tailor the social organization of society to the ecological requirements of the life support system.

The Eastern world view departs radically from that of the West regarding man's relationship to nature. The domination of nature has never been a primary occupation of the Chinese people, as it has of Europeans or Americans. Economic efficiency has always involved the concern for human welfare, with the conviction that such a process in the long run will indeed turn out to be the most efficient path to a society of free and creative people. The Chinese are certainly not without a record of ecological or social disorder. Great contradictions certainly exist within China, not least of which is the hierarchical structure of leadership or the production of atomic weapons. But given the almost overnight emergence of Chinese society from feudalism and the present competition within the socialist bloc between China and the Soviet Union, not to mention the twenty years of red-baiting that has consumed the United States, one can understand current contradictions in terms of long-range objectives.

What is most critical to note about the Maoist model for socialist development is the convergence of social motivation and ecological balance, such that the logic of the social order is tailored to compliment the evolution of the life support system, not destroy it. This model seeks to integrate political with ecological jurisdictions, not to separate them.

This example is illustrative of the different implications between capitalist and socialist development. In a future work in preparation I will explore the ecology of socialism. But at this point, several critical issues are apparent: while capitalism promotes the profit of the individual, socialism cherishes the collective welfare of all; while capitalism promotes competition as the basis of the social contract, socialism seeks the path of cooperation; while capitalism depends upon ceaseless growth in terms of production and consumption, socialism works toward a balance between production and human creativity. The Fanshen[7] experience of China is only one illustration of a variety of socialist experiments in collectivization throughout Asia and Europe.

It is impossible here to do anything but suggest what is happening in several parts of the world. More important than any of the particulars is the overwhelming confirmation such experiments provide about the ability of people to radically transform their life pattern.

iii

The practice of the twentieth century is rich with the experiment of self-determination and workers' control. While the nineteenth century, except for the Paris Commune, was a time for theory, ours is a time for practice, leading to a higher level of theory. Given the variety of experiments, little integrated information is available.[8]

The experiments of Europe, including Israel, have three basic characteristics. The first can be seen in the communautés des travails which developed in France subsequent to World War II. Essentially these forms developed a cooperative ownership of the industrial plant and the democratic participation of workers in the management of production. In Yugoslavia, the forms of workers' self-management, while often not as adequate in terms of self-reliance as the term suggests, have provided valuable lessons regarding the different levels of government control and the process of self-government.

Within both the French and the Yugoslav models the most important shortcoming has been the failure to fully integrate the ecological, political, and economic aspects of social life. In Israel, within the kibbutz and moshave, this has been attempted on a magnitude achieved nowhere else in the Western world. This interplay is critical to several of the more significant issues which face such microsocialist experiments. Hierarchy between political, economic, and ecological considerations can only be abolished once these jurisdictions begin to blur, begin to synthesize into an entirely authentic form of self-rule, where political and biological principles are considered interdependent. Self-determination will require, in a decentralized context, an adequate and mutually supportive network of basic skills and services which promote the maximum diversification of function, role, and participation in the collective struggle. The prior example of "multiple use" in China illustrated this concern.

The Asian models of decentralization in North Vietnam, North Korea, or China are motivated by a relatively low level of industrialization as compared to the West. While the scale of their experiment is quite unlike that of the United States in terms of industrial output, the forms of socialist industrial management they have created include methods applicable to any industrial situation. Moreover they have demonstrated that decentralization and self-reliance, rather than being a burden to production, can surpass former levels of centralized productive capacity, because production is considered an ideological, not mechanical, task.

The decentralization of North Vietnam[9] and the evacuation and dispersal of its population in the face of American air power constitute a remarkable lesson in the flexibility and strength of a people to drastically alter their form of life when survival is at stake. All indications are that the North Vietnamese will continue to pursue a decentralized and diversified form of development after the Americans have left Indochina.

China, Cuba, and North Korea differ fundamentally from North Vietnam. We can observe here the struggle of a revolutionary society to build a communist social fabric after the armed struggle against American imperialism succeeded. While Cuba and China are essentially agricultural societies and will remain so for some time, the Democratic People's Republic of Korea has, since the war, achieved an unprecedented level of industrialization and development for a Third World nation. Perhaps most significant is the lesson North Korea offers about the importance of authentic self-determination as a path to socialism for small Third World nations.

An example which well demonstrates the mechanisms of self-determination, was the reported unwillingness of North Koreans to mass produce automobiles though the Soviet Union offered them resources. While information about Korea is limited, we might well speculate on the course of a Communist society when confronted with such alternatives. Aside from the ecological effects of the auto, more basic human and social considerations come into play. What would the privatization of transportation do to the social fabric? How would it affect the level and frequency of social intercourse? As a result of a social structure where the primary concern was social responsibility and collective participation in basic services, it might be determined that the auto would actually encourage attitudes and behavior that were unhealthy to the socialist development of the country. This process of selection in regard to transportation is motivated by a spirit of collective struggle and mutual aid and by a constant concern for the interplay between individual and social, rural and urban, private and public. On the deepest level it is motivated by the common purpose and will of the Korean people, and their desire to preserve a society built on trust and cooperation, rather than competition.

The achievement of the North Korean people[10] in rebuilding their country after the American war against Korea is a potent lesson in the viability and superiority of socialist development — particularly when compared to the postwar development of South Korea.

Today the Democratic People's Republic of Korea (DPRK) is the most highly industrialized of all socialist nations in Asia or anywhere among the small Third World nations. The popular American conception of Korea as a grim, cold, and autocratic nation is nothing but a vestige of American anticommunism, and is sharply dispelled by anyone who visits North Korea. Travelers throughout socialist countries speak of North Korea as among the most beautiful, even though Americans after the war were fond of boasting that "not a single tree is left standing in North Korea."

The juche[11] principle of self-reliance promoted by Kim Il Sung, the Premier of North Korea, has led the DPRK to create a balanced independent national economy. Nowhere in the socialist world has the principle of self-reliance been so thoroughly developed as in North Korea. In order to achieve a balanced growth, the North

Koreans have given priority to the development of heavy industry while at the same time proceeding with light industry and agriculture. Great concern has been exercised to maximize the symbiotic relationship between agriculture and industry, rural and urban areas, peasant and worker. The unique combination of accumulation and consumption within the economy has allowed a distribution of national income which both provides for a stable economic foundation as well as systematically raises the standard of living. Self-reliance has led to the domestic production of practically all consumer and manufacturing goods, half of which are produced on a local scale. While Korean exports compete favorably in European markets, foreign assistance and cooperation come secondary to sustaining national independence.

The DPRK has pioneered in the creation of many unique forms of socialist management. Kim Il Sung spends 80 percent of his time traveling throughout the country. As a consequence very many North Koreans have either talked with Kim Il Sung or seen him personally. The Chongsanri Spirit and Method and the Adean system of work management are exemplary forms of proletarian communism.[12]

Perhaps most important about the DPRK is the extent to which economic progress has been matched by ideological transformation. In all forms of production, precedence is given to political work. With the highest standard of living of any Asian socialist nation, the DPRK has also succeeded in bringing one out of every five eligible adults into the Party — the highest proportion of Party membership of any communist country.

The examples cited from the Soviet Union, China, and North Korea can only begin to illustrate the basic ecological stability of socialist development.

iv

The source of balance between politics and biology within Communist society stems from the preeminent sense of value given to human development. As a consequence, production is tailored to maximize human welfare, not human welfare to maximize production. Within such a context clear and discernible limits are set on production, limits which determine the social parameters of life.

The social balance which is the precondition for the equilibrium between social and natural fabrics must tailor its organization along similar principles as the evolutionary process itself. This must involve fundamental reconsiderations about the role and purposes of competition in the social framework, about cooperation and symbiosis and the entire normative structure of capitalist society, where consumption is determined by dollar value rather than natural capability.

Balance is always concerned with limits, with boundaries, familiar guidelines. Without such rationality balance is impossible. Otherwise the center, as well as the limitations, are always in question. America epitomizes the loss of center, the loss of common goals and aspirations which Martin Buber characterizes as the crux of community.[11]

In the natural world balance results from the conditioning of animals and their ecosystems, so that populations do not exceed the resource-carrying capacity of the natural system. Balance is maintained through the natural limitations of energy and population. If a species exceeds in numbers the available energy budget in a region, that species will perish, or fall subject to the natural limitations of the new ecosystem into which it moves.

In the social dimension, capitalist society has long ago surpassed the carrying capacity of the life support system. The extreme pressure brought to bear upon natural systems is manifested by the smog around Los Angeles. There simply are not sufficient atmospheric conditions to absorb the amounts of contaminants poured into the sky from automobiles. The same principle applies to the oceans, rivers, and the nutrient value of lands. The unlimited production and consumption of commodities and hazardous chemical agents under capitalism is rooted in the ideology of the economic structure, the mechanisms of the market and the competitive nature of society.

Are not socialist countries concerned with growth? They certainly are. Economic development is a primary concern of developing socialist nations. But the question is not simply one of growth or no growth. Growth is the basic pattern of the universe, nothing is static. The talk of a "no-growth" society in the United States is ludicrous. The question is what kinds of growth must be encouraged and what kinds of growth should be prohibited, not whether growth itself should be condemned wholesale. Flowers grow, children grow, personalities grow. So can capital, production, and consumption.

Capitalism, motivated by private profit and competition, sets no rational limits to the production of commodities, save those of capital profitability. This is not the case under socialism. Economic growth is a critical concern of all socialist societies. This is because socialist societies still operate under conditions of scarcity. Centralized planning and coordination are required in order to eliminate the conditions of deprivation. Much of Marxist thought today clings to that historical period in which Marx formulated his original teachings, without realizing in fact the dawning of conditions which must of necessity temper the contemporary Marxist view of history.[13] For our purposes, at the center of this tension is the Marxist view of man's position in nature. There can be no denying that for Marx, man was the measure of all things. Marx believed that the inevitable progression of Communism was from nature's domination over man, to man's domination over man, and finally, under Communism, to man's domination over nature. This vision entails a view of historical development which could not foresee the possibility that the realm of necessity might be incorporated into the realm of freedom within the developed capitalist world.[14]

Marcuse notes that Marx believed, in essence, that "only those who were free from the blessings of capitalism could possibly change it into a free society: those whose existence was the very negation of capitalist property could become the historical agents of liberation."[15] In other words, Marx did not envision a situation in which the necessity of scarcity might be eliminated before the elimination of capitalism. In such a state, what Marx argued against Fourier, becomes in reality, that work would be transformed into play. Such a possibility is today evident with the advent of a liberating technology[16] in which the conditions which for thousands of years required alienating labor might be abolished. Marcuse has developed this possibility in "The End of Utopia":[17] the very "technification of domination undermines the foundation of domination," or that "the path to socialism may proceed from science to utopia and not from utopia to science."

Such a historical reality would call into question many of the most fundamental conceptions of contemporary Marxism — a theory predicated on and propagated under the assumption that the hierarchical nature of human society would be an inevitable necessity given the conditions of scarcity, such that the road to Communism would require the hierarchical organization of "a dictatorship of the proletariat."

In his time Marx did not anticipate the extent to which capital would be centralized and the more recent developments of cybernetics and technology. These events which dominate the history of our age challenge the Marxist view that power must be transferred from one class to another, from the bourgeoisie to the proletariat which somehow would lead to the withering away of the state. The maintenance of the power nexus was rooted in a view of history which did not anticipate a situation in which scarcity, the very justification for that hierarchy, might be eliminated before the arrival of Communism — the abolition of both private property and alienation.

The technological innovation of the present age has the most profound effect upon traditional class distinctions. The necessity of class struggle was formulated in an age when the primary form of colonization was the individual relationship to the means of production. The conditions of the nineteenth century focused attention upon the harsh inequities of the industrial revolution. Today, production for the sake of production finds its complement in consumption for the sake of consumption. The present level of material abundance within a large sector of capitalist society has today incorporated the very demands put forth in the *Communist Manifesto*.

The traditional proposition of class as the agent of transition from a society of class to a classless society anticipates neither the effect of affluence and consumption in generating revolutionary consciousness nor the extent to which the division of labor itself would proceed — "aside from race and sex the industrial work force is stratified along lines of age, skills, 'mental' vs. 'manual' salary differential, level of authority, status, education, etc."[18] Such stratification raises questions about current possibilities for a majoritarian movement within the proletariat of industrial capitalism which does not at the same time "raise issues of racism, sexism and stratification both as autonomous issues and as integral to the oppression of the working class as a whole. It would have to raise these issues among all workers and throughout society, as well as within industrial production."[19]

This is not to deny the historical importance of the proletariat, but to root the consciousness of the proletariat within our own historical moment. For nowhere is the cost of the capitalist mode of production more blatant than in the factory itself. The advent of industrialization ushers in a new era of exploitation and human misery. The cost of the affluent life — to those who produce it — is staggering.

Findings of a study in 1968 conducted by the Chicago Institute of Medicine revealed that 73 percent of Chicago's work places exposed workers to one or more potentially hazardous materials such as toxic solids, liquids, or gases. The same year, Dr. William H. Stewart, the Surgeon General of the United States, told Congress that studies by the Public Health Service revealed that 65 percent of industrial workers are exposed to toxic materials or harmful agents such as severe noise or vibration. He found that only 25 percent of the workers were safeguarded *in any way* from these hazards.

While we are publicly reminded about the numbers of people killed in Indochina or the figures for those who die in automobile accidents on every holiday, we are not reminded, if told at all, about the systematic assault on men and women who are killed, maimed, mutilated, or infected every day on the job.

The magnitude of this assault is grasped when we realize that the health director of Sweden's largest Volvo plant, at Gothenburg, "strongly advises workers not to spend more than two years of their lives on the production line. After two years, they will be threatened with lasting mental or nervous damage that may disable them for the rest of their lives."

Similar conditions obviously exist in virtually all assembly lines and all forms of mass production. Ray Davidson has chronicled in *Perils on the Job* the daily physical assault against workers. Every year there are 14,000 industrial deaths with another 390,000 people contracting various industrial diseases from mercury, lead, arsenic, carbon dioxide, benzene, asbestos, and the like. . . . Every day 8,500 workers are disabled and another 27,000 injured in some way. Davidson notes "that the records are tragically incomplete."[20]

A Special Report issued in 1966 by the Surgeon General, "Protecting the Health of 80 Million Americans," stated that every twenty minutes a new and potentially toxic chemical is introduced into industry. Estimates today hover around 450,000 different chemicals and industrial materials in use. Research into the short-term effect of such agents upon workers exists for several hundred. Long-term chronic effects are ignored, to say nothing of the synergistic effects of two or more of such agents working together.

Today in the United States there are nearly twice as many game wardens as there are industrial health safety inspectors. The daily crime committed against the 80 million workers of America proceeds apace, government more concerned about crime in the streets than crime on the job.

The unrest among workers characteristic of the last few years finds its expression in municipal strikes, slowdowns, work stoppages, and wildcats of all varieties. Perhaps the best organized and most important of all working class developments is the League of Revolutionary Black Workers in Detroit.[21] To date, workers' struggles are predominantly underclass, led by black workers.

By and large the experiments among workers in capitalist society today are concerned with the preconditions of freedom. They are the beginning of a revolutionary process which will transform not only the means of production but also the concept of production itself. The conditions of hierarchy and scarcity dominate the form of social organization; this is the precondition of freedom. It is the condition of all underclass people in the industrial world.

Yet today throughout the capitalist world the realm of freedom — freedom from scarcity — constitutes an entirely new historical project, connecting the events of May-June 1968 in France to the People's Park. At such moments liberation becomes not the object of struggle but the struggle itself. The imagination of rebellion replaces the illusion of philanthropy. It is a new dawning of the classical Marxist notion that socialist revolution would occur in the most developed of capitalist countries. Marcuse recognized this in "the utopian concept of socialism which envisages the ingression of freedom into the realm of necessity, and the union between causality by necessity and causality by freedom. The first would mean passing from Marx to Fourier, the second from realism to surrealism."[22]

To pass from production as labor to production as play under capitalism requires the emergence of qualitatively authentic human needs, needs which appear, as Marcuse and Bookchin have pointed out, as biological necessities. The massive opposition within the fabric of American life, opposition within universities, military, unions; the rejection of the affluent life reveals a process of disintegration which is widespread. This disintegration reaches into the family and work place. It raises new issues, new forms of struggle, and will call for new responses to the questions of organization and praxis.

Perhaps in miniature the People's Park in Berkeley was a step in the United States analogous to the events of May-June in France. The struggle to build and defend the People's Park in 1969 was the result of the continual incursion upon the youth culture of Berkeley by the business community's plans for modernization. Brown shingle houses were systematically torn down to make way for plastic apartments. Street commerce was converted into department stores. Open spaces were devoured by parking lots. This constituted a microcosm of the Bay area region, in which twenty-one square miles of open space were being devoured by asphalt and development every six months — an area half the size of the city and county of San Francisco. Then, spontaneously, a group of people formed in Berkeley and began to convert an open lot, occasionally used for parking, into a much-needed park. This land was owned by the university and destined to be converted into high-rise dormitories, even though the present dormitories were partly empty. The creation of the People's Park involved people from various sectors of the community. It affirmed the need for mutual aid, participation in creative, communal activity, open space, room for children in a crowded community of automobiles, and most important, an opportunity for self-definition in the face of the continued exploitation of the Berkeley neighborhoods. The park represented the necessity of people to design a community which reflected the organic needs and purposes of the people who lived there. This effort sought to reassert balance in the Berkeley community, between people and asphalt, open space and congestion, quiet and noise, diversity and monotony.

The worldwide affirmation of this momentum created in subsequent weeks People's Parks in many parts of the United States and Europe. More than all the subsequent ecology teach-ins, legislation and the like, the spirit behind the People's Park represented a motor force behind the contemporary concern about ecological imbalance. More important, it provided a model for the integration of social and ecological necessity. It provided a lesson as well, as to the kind of force which will be applied when people come together to express their affirmation for a new culture.

Since the People's Park, the Berkeley community has been converted into a vast experiment in self-determination. A group known as the People's Architects began to initiate a community design process for land being vacated for mass transit construction. Numerous food and service cooperatives have been developed. The Food Conspiracy now claims to serve between three and six thousand people. More recently, an initiative for the community control of the police was put on the ballot.

The People's Park is not an isolated event, nor simply symptomatic of Berkeley. It must be seen within a larger context of a variety of community development and self-determination experiments currently apace within the United States and other countries. Such movements within the United States suggest the reconstitution of limits, of boundaries. This will not and cannot be achieved through a centralized mechanism, whether new Federal regulation or a national party. Further centralization today can only serve to further destroy the limited natural and social diversity which remains.

What has been assembled over decades, if not centuries, will not be transformed overnight. The gravity and extremely fragile nature of our present condition must suggest the tentative conditions of any potential solutions. Massive programs and legislation or immediate demolition are both as inadequate as the other. Solutions that can today be imagined must be suggested not as blueprints, not as final solutions, but as fragments of an initial healing process. There is the most fundamental difference between the attempt to rectify, to patch up or repair the effects or defects of centralized production and personality, the results of decades of hierarchical organization in industry and social intercourse, and the generation of the first of many projects in a revolutionary process, animated by a symbiotic vision of a society steadily integrated into the natural fabric in a way mutually beneficial to men, women, and all other forms of life on this planet.

The present conditions of social and biological upheaval result from the application of massive programs and centralized technology, from vast bureaucracies and impersonal proposals. More of the same, more Federal government, more centralized technology, more academic research — more of the same potion, will not cure a disease which today feeds on its self-propelled solutions.

It is hard to imagine an alternative course. We are taught to think of revolution as a turn of the switch, a massive and sudden upheaval, as an overnight affair. Immediately the cultural organs of the *Reader's Digest* mentality provide images of the Soviet Union or Cuba. In rare cases, even China. And yet such nations underwent various forms of revolutionary upheaval under social and economic conditions which bear not the slightest resemblance to those of present-day America.

What is at stake today everywhere is not simply the transfer of power from the hands of the present few to the hands of another few. We must be concerned with the elimination of the power nexus itself, in all its everyday manifestations.

The prospect of such a movement depends upon nothing as much as what Andre Gorz has called the formulation of a "total alternative" — an outline to an entirely authentic model of production and consumption rooted in and motivated by democratically determined needs, on a scale capable of being defined in part by each of its participants, but together as well, as a whole.

Yet much of recent history constitutes a rush from alternatives. Only recently within America has the concern for something better, for a living, concrete alternative, been present. By and large, it has been a "century which moved with the magnificent display of power into directions it could not comprehend. The itch was to accelerate — the metaphysical direction unknown." Nothing demonstrates this more than the simple truth that today the popular media devote more attention to the moon than they do to the exploration and understanding of local government. Many Americans know more about the surface of the moon and the Apollo missions than they do about the workings of their own neighborhood. It is increasingly easier for us to get information from faraway places than from across the street, or from the person we may be living with. While the human horizon may be expanding, human beings have become narrower.

It was in large measure this realization which was the source of an entire decade of protest during the 1960's. R. D. Laing summarized this sentiment "At all events, we are bemused and crazed creatures, strangers to our true selves, to one another, and to the spiritual and material world. . . . We are born into a world where alienation awaits us. We are potentially men, but are in an alienated state, and this state is not simply a natural system. Alienation as our present destiny is achieved only by outrageous violence perpetrated by human beings on human beings."

The recognition of this reality began when northern blacks and whites went South to register voters. From there, young white students began to organize in the poor ghettos of the North. The lessons of the 1960's were in the streets and through the media, not in the classroom. The lessons ranged from Watts to Birmingham, from Cuba and the Dominican Republic to Vietnam and Indochina. The amplified inequities and injustices of racism, poverty, and war demonstrated time and time again that the promise of America as a land of plenty, a peace-loving nation was a cruel hoax, at home and abroad.

It was more than anything this opening of perceptual space which is today at the source of the present understanding about the social and ecological imbalance which threatens the very continuation of life on this planet. Through the sixties the American dream was up against the wall. Today, it is no longer a question of growing up absurd, but of not growing up at all.

We must recognize the assertion of limits, whether it be on the production of lethal chemicals or the local control of police, both are inseparably crucial experiments in the refabrication of an ecologically sound social order.

The quest for limits found its origins in the early sixties with the slogan, "Let the people decide." The struggle for social diversity, whether in regard to the composition of a lunch counter or the determination of a local community to partake in the utilization of resources, is the basis for natural diversity. Many of the motivations which move people in other parts of the world to define self-government today infuse experiments in countless forms, both rural and urban, which dot the American landscape. There is a long history of communal experimentation in the American grain. From 1780 to 1860 one hundred American utopian communities were founded. Less than a dozen of them survived more than fifteen years. And yet today, such developments as communes, collectives, food cooperatives, craft guilds, and rural-urban collectives function in and around almost every large city in the United States. The *New York Times* of December 17, 1970, counted nearly 2,000 communities in some thirty-four states. It concluded that the figure was conservative, and no doubt it is.

To these experiments must be added the growth of community development corporations around the country and a vast variety of community and neighborhood organizations. While such experiments take many forms and dimensions, ranging from new black rural towns to hippie enclaves and community development corporations in the ghetto, all share a common and overriding motivation — the struggle for self-determination, the reassertion of limits and self-reliance. Such experiments deal with a vast variety of social issues and problems, ranging from the structure of the family to neighborhood health and education as well as self-defense. Some are more dependent upon the demands of capitalist society than others. Some openly reject society as it exists now, others seek merely to gain their "deserved" share of the pie.

Few have considered seriously the implications of autonomy and self-rule in its full dimension or begun to speculate about the implications of such developments for new forms of government and association, the possibilities for authentic forms of technology and the proper limitations in regard to scale and diversity. Few have seriously contemplated the full interdependence between political, economic, and ecological requirements. And most important, few could have imagined but a decade ago the remarkable profusion of such sentiment and experimentation.

Inherent within these projects are two motor forces, decentralization and diversification. Our present cities began as small towns and villages. Gradually they were consolidated under a central or regional network. The government of neighborhoods was abolished, and thus the autonomy of style and ethnic origin, though of course this has persisted.

The result of this centralization, as Milton Kotler has demonstrated in *Neighborhood Government,*[23] is the impoverishment of political life. The overriding consequence of centralization has been the elimination of diversity. The monotony of centralization, as in the army or the factory, is rejected in favor of increasing self-reliance and social choice in regard to basic necessities as well as luxuries.

There obviously exists today a variety of opinion regarding the proper strategy to achieve self-reliance. Kotler believes that "the object of local power can be nothing less than re-creating neighborhood government which has political autonomy and representation in larger units. . . .[24] But the democratic aim of local liberty does not depend upon a change in the national power structure. Local control must be the preliminary fact and remain the final object of political revolution."[25]

Here is the critical question about social and biological liberation. Can balance be reconstituted on a purely local level? Can ecological and social costs of the nation-state as we know it be permitted to exist? Can the legal incorporation of local territory, such as a neighborhood, provide for the transfer of adequate public authority and autonomy to achieve self-determination? Kotler and others believe it can. But here is the crux of the interdependence between political and biological concerns. The example that Kotler has nourished is ECCO, the East Central

Citizens Organization in Columbus, Ohio. This project has served as a model for subsequent organizations in several cities around the country. It has succeeded in acquiring important forms of autonomy in the performance of services for the community. Particularly in poor and underclass communities, the importance of basic necessities must not be understated.

But we must realize as well that the overriding practice of capitalist society will be to promise both bread and circuses and in the end, provide only the circuses. There is no way to transform the required productive capacity of the United States in order to provide for basic services without presenting a challenge to the organized mechanisms of centralization which determine productive priorities. Such a challenge cannot overlook the critical importance of threatening state power.

It is here that biological and political interdependence becomes most pronounced. For no matter what autonomy a local community may achieve in regard to services, perhaps police power and even, in time, capital outflow, or taxes, these victories have not brought with them the right of biological self-determination. As long as a productive apparatus exists which continues to produce anything like the present level of commodities, predicated upon increasing profit and thus increasing waste, the biological balance of all people and all living communities will be threatened. The mercury or lead produced in one continent will work its way into every living community. That is why the productive capacity of the state, the state itself, must be challenged, and, in time, dispersed into wholly new forms of self-rule on entirely unique scales and dimensions.

The dimensions of political government must be tailored exactly to the carrying capacity of the ecosystem in which it exists. The parameters of this approach have already been sketched, by Lewis Mumford in *Technics and Civilization*, and Ian McHarg, in *Design with Nature.*

But here again the prerequisite for such ecological homework is the established authority and legitimacy of the neighborhood or multineighborhood unit. This level of government must secure the power and capacity to determine such critical issues as the location and forms of transportation, the agricultural and energy capacities, and desirable locations for settlement and production in accord with geology, hydrology, and other natural conditions. Settlements in this way can be tailored to the nascent logic of the landscape, in accord with those levels of consumption and production which the natural fabric can sustain. The obvious result will be a vast diversity of consumptive levels and a natural life style, predicated by the available food supplies, energy, and other life-sustaining systems.

The foundation of such units must originate in the neighborhood, an area Kotler has described as between one-half square mile and six square miles, defined primarily by physical boundaries and barriers which naturally intensify local relationships. Already today throughout the United States there are some seventy-five neighborhood corporations.

It is obvious that such neighborhoods will not have the required resources for complete self-sufficiency. But the natural ecosystem in which they are situated, the San Francisco Bay Area, New England, Appalachia, for example, will have adequate resources in most cases. The rule should be self-sufficiency where possible, so as to minimize, over time, the necessity of centralized control and distribution of natural resources and energy. Political and ecological jurisdiction must seek direct interface at the lowest common denominator, expanding organically.

There are obviously different functions which are appropriate to different scales. No doubt environmental regulation and transportation will require regional coordination, whereas energy, education, health, and other services may not; here again it will depend upon the particular locale, the interests expressed by the people, and the natural limitations of the area,

The object is to maximize self-reliance in regard to the operating principles of the ecosystem. Political boundaries must be tailored to reflect both social choice and ecological conditions. Today they reflect little more than the concentration of capital. From such a perspective, many forms of government, particularly the nation-state within the overdeveloped world, would be eliminated.

vi

Increasingly Federal and state government serve no authentic function in regard to people's actual needs. The political and economic authority they have acquired has drained the neighborhood of its vitality and legitimacy, because they in fact drain the neighborhood of its resources: human, capital, cultural, and otherwise. Few recognize the extent to which the state drains local communities of their vital resources. This in large measure explains the fiscal crisis of the cities. Take, as an example, New York. The skyrocketing city budget illustrates the problem:

Major NYC Expenditures

(Billions of dollars)	1966	1971
Hospitals	$0.269	$0.539
Police, Fire and Sanitation	$0.667	$1.200
Welfare	$0.494	$1.712
Education	$0.857	$1.784

SOURCE: *New York Times,* March 7, 1971.

At the same time, the people of New York pay out in taxes five dollars for every dollar which returns. New York City pays out $14 billion in income tax to state and Federal government; back comes $3 billion. As Norman Mailer described it, "In relation to the Federal government, the city is like a sharecropper who lives forever in debt of the company store."[26] New towns or cities, given the present structure of capital and taxation, do not escape this condition.

The Federal colonization of cities in regard to capital and resources explains something about the current urban crisis, particularly the collision between cities and unions. Within the past decade government-employee unions alone doubled in membership to close to three million persons. The growth of union membership among government employees grows faster than in any other employee groups at the same time that government employment is the largest expanding employment sector. In 1959 there were twenty-six strikes in which 2,240 state and local government workers participated. In 1969, there were 409 strikes involving 159,400 employees. In 1970 there was a municipal workers strike almost every three days in the United States. This condition results in large measure from the fact that the money for wage increases demanded by employees is spent in Indochina.

Consequently, the tax structure of major urban areas is one critical arena for struggle toward self-determination. While neighborhood organization may achieve some autonomy over services, political or ecological self-determination must begin with control over the purse. Capital cannot continue to flow out of urban areas at the present rate and expect to undertake anything like the level of reconstruction which is required. As pointed out earlier, the vast inequities in taxation are an excellent inducement for opposition. Taxation and control over capital must be one central focus of any political movement for self-determination.

Part of the reason that tax revolts have not been very successful in recent years is the individual nature of opposition. It is irrelevant for an individual to merely withhold his taxes. More important, there is nothing to which that money is allocated in an affirmative manner which enforces opposition. But once political jurisdiction is established on a neighborhood or transneighborhood level, taxes could begin to be funneled into such political entities rather than the Federal government. There is no illusion here about the consequences of such an act, but equally there must be no illusion that we can continue to pay taxes to state and Federal government on the present scale and proceed with any form of creative economic development on a local level.

There are a number of other steps which local political organizations can begin to implement once authority is established among its members. Many of these proposals can be raised on a national level as well. But it is important to point out that such proposals are only of educational value when raised nationally. Implementation requires political power, power that today can only be achieved on a concrete local basis.

A notion which might be explored immediately is the question of rationing on a community-wide basis. While the threat of rationing has been raised publicly in regard to national fuel shortages, it is always proposed from the top down, and thus is merely a matter of those at the top determining priorities. In such a context kilowattage would be cut off to ghetto areas and channeled for police and other administrative purposes in times of shortage. Such programs serve merely to bolster the current distribution of political power, and constitute not so much rationing as a method of preserving the current distribution of resources.

However, rationing on a community basis entails something else. To begin with, political dependence locally will require a community to establish contact with its material foundation. In other words, in a society of excessive surplus production, value tends to be determined by highly artificial forces, such as advertising, taste, and style. Rationing might become a method for a community to assert limitations on local production and consumption which are geared to the wealth and values of that particular community, rather than a national marketing system.

Within this context rationing would serve to reassert limitations between the production and consumption of commodities and the local carrying capacity of the ecosystem. Natural resources, including oxygen, water, and land, could be rationed by a community for designated purposes, rather than remain subject to market mechanisms.

But perhaps most important, such a notion would entail the widespread participation of a community in coming to understand the working mechanisms of social organization and the individual and collective relationships involved. People here would begin to consider the origin and purpose of basic services now taken for granted, such as energy, transportation, the production of food, the utilization of communications, the distribution and quality of health and education services, for a beginning. Such a process would help to break down the extenuated specialization which today is a fundamental source of powerlessness amongst people — the lack of concrete knowledge about how things work. Such knowledge is the prerequisite for understanding dependence and moving toward liberation. There is no reason why a neighborhood of a few thousand people or a cooperative of tens of thousands could not rationally proceed to make such decisions within the current situation. Such practice would lead inevitably to valuable knowledge about the limits and possibilities of self-determination today, how political power is constituted, and what is entailed to break the chain of domination.

Inherent within the proposal for democratic rationing is the attempt to tailor the political organization of society to the nascent logic of the landscape. In so doing, one cannot escape considering the entire mode of capitalist production. Today it is clear that the index of productive capacity is no measure of quality of life. With the highest level of production in the world, we have the worst system of transportation of any industrial nation. Today in the United States any serious improvement in the quality of life may well depend upon the initial reduction of productive output. Cutting back production through the procedure of community control could be an important mechanism for identifying the extent of surplus production in a capitalist economy. A transition to a socialist economy will depend upon the ability to gear production to democratically determined needs, not market maneuverings.

Production geared to profit inherently determines production patterns in which the maximum added value is attached to everything produced. Use value is minimal. For instance, the energy and materials required for the individual motor vehicle, as recognized earlier, are the major underpinnings of the American economy. And yet, such a system of transport verges on total nonfunction today. A balanced, diversified system of public transport, determined by rational patterns of population distribution, could operate at a marked reduction of the current production costs as well as provide considerably better transportation than the present individual mode. The force required to move the automobile serves to move the hulk of the vehicle, not the person. Moreover, the amount of energy expended to move individuals rather than collectives is a remarkable waste. If one were to consider the energy and resources conserved within the context of a rational population distribution and a diversified transport system, somewhere between 50 and 70 percent of the current cost of transportation could be reduced. This would include steel, plastics, highways, gasoline stations, et cetera.

Today there is virtually no justification for producing another automobile. Nonpolluting engines could be installed in current frames. Considering that auto sales account for 24 percent of retail sales annually, one gets a picture of the conservation that could be achieved. The first question to arise is how do the people currently employed obtain jobs if the automobile is cut out. Obviously, such a step could only occur in the context of other fundamental changes, including the redistribution of wealth and resources. People could then be employed by their own communities, to rebuild the fabric of urban life if taxes were not drained by a Federal government. In short, one step cannot be considered without the others.

Other forms of production can be considered. Most appliances and electrical equipment are produced to serve highly specialized functions. In large measure, this results from the high proportion of added value associated with such a form of specialization. However, it would be considerably less costly to produce multiple-purpose appliances and machinery for needs determined by local democratic control over the means of production, rather than a myriad of appliances and gadgets produced today which either serve no real function or are packaged in such a manner as to upgrade the actual cost by several hundred percent. The extent of production geared to absolutely useless products or products actually harmful was noted earlier in our discussion of advertising.

The single family suburban household is a microcosm for this irrationality. Consider what transpires in each of the various rooms of the house, the extent of wasted space, or energy. Consider as well the function of the household in fulfilling desired social roles, the separation from each other and those in one's neighborhood, and the picture begins to fill out. Take the adult bedroom. While children in fact actually require more space, and utilize that space more, they are always given the smaller bedrooms. Adults have no use for the bedroom except as a place, in private, to make love. Affection outside the bedroom is taboo. People sleep in the bedroom but it is without use, by and large, for the remaining hours of the day. Many Asian peoples simply pull out a bed when they want to sleep. Consider as well, the fulfillment of sexual roles exhibited by the various rooms of the house. The kitchen is the woman's room, the garage or study the man's. Women cook, men are active mechanically or intellectually.

Most of us hardly stop to consider the extent to which such social organization is both destructive to social relations — a means of preserving particular sexual roles, a form of separation from other people — and highly wasteful. Take for instance the driveway, garage, and yard. It is obvious that better than each of us having a tiny plot of land, a block or neighborhood or a number of houses could combine their land and create a minipark — a far better place for children than the isolation of the individual yard. The pooling of such resources could lead to child care facilities, adequate toys, and a variety of educational experiences which the children would never receive in isolation. Such a step would move as well toward the collective responsibility for child care.

The space required for everyone to have his own garage, his own driveway, his own yard, is quite an extravagance.

The household is a microcosm of alienation and waste for the neighborhood, city region, and nation as a whole. Separated and atomized as individuals, we are helpless to overcome this spiral of exploitation.

To this picture must be added the extent to which production in America is geared exclusively to death — defense spending, cigarettes, and the vast variety of products which are either unsafe or outright lethal. Together this accounting of surplus and exploitation may amount to somewhere on the order of 60 to 80 percent of the current output. In other words, given a humane and rational social organization of human and material resources, a far better quality of life could be sustained in the United States for 20 to 30 percent the present expenditure of energy and resources. If this seems hard to believe, I suggest that one merely begin to examine his own living situation in this perspective. Take a large piece of paper, lay it out on the floor, and sketch a floor plan of the house. Bring in your neighbors, and begin to examine room by room what happens there. What people do, what resources are utilized for what purposes, what kind of energy is consumed, what social roles are implied, and the picture will begin to fill out. Then extend such considerations to your block, neighborhood, city, et cetera.

A similar model could be applied to the entire capitalist world, including Japan, West Germany, France.... The yield in potential resources and energy would be staggering — a yield which could be put to the task of global social reconstruction. A good part of that yield would simply involve leaving the world's resources where they are — in the developing nations.

Such transformation in our everyday life could not occur without a profound cultural revolution in the United States, one in which the competitive, individualistic attitudes and values which today dominate our lives were organically replaced by more cooperative and collective identification with the common welfare. While this cannot happen overnight, one should not underestimate the opportunity for collective life that is today available for those who seek it.

A second factor to consider is the form of social organization in developing countries which could best provide for the health and welfare of the population. We already noted the progression of China in a period of more than twenty years from a situation verging on mass famine and starvation to one of adequate capability to care for the welfare of the population. Here are one-quarter of the people of the world. What accounted for the fundamental transformation? More technology? More food? Yes, both are involved. In the last five years China has produced a bumper crop of rice. But what accounts for the massive disparity between India and China? I would contend basically that it is the social organization and motivation of the Chinese people which is at the root of this disparity, an organization in which production was increased and population leveled, in which egalitarian distribution and social cohesion were achieved because of Communism.

There are many other tactics which might be adopted which could tie together existing sentiment to local political power. One such issue is the workweek. At present there are over one hundred companies nationwide which are experimenting with a four-day week and a ten-hour day. Some are actually developing a three-day week and a twelve-hour day. Labor argues in this context for spreading out employment, that is for less hours and higher wages. Could such proposals on the part of unions be tied to the needs for community development? Could part of the bargaining package of a union involve a stipulation that companies situated in a defined community should curtail the working hours of production, maintain the present wage scale, and contribute those hours in some form to a community development corporation in which people began to be paid by local businessmen for the reconstruction of the local community? In other words, could a method be found through labor to require a business to invest part of its earnings back into the welfare of the community in which it is situated? This is one example of scores which could be negotiated through labor contracts. Already some unions have proposed and secured pollution limitations as part of the bargaining contract.

Many such proposals today could be a rallying point around which a number of local or community groups could begin to consider primitive forms of cooperation or confederation into regional bodies. The consciousness of this necessity is already widespread among certain sectors of the society, and perhaps received its most publicized attention in the 1969 New York mayoral campaign of Norman Mailer and Jimmy Breslin where they campaigned around power to the neighborhoods and New York City as a fifty-first state. Similar sentiments today motivate much of the activity under the banner of the Vermont Free Territory or the initial regional organizing attempts which are occurring in New England, Appalachia, or the San Francisco Bay Area. One of the most interesting and highly developed urban alternatives is Le Front D'Action Politique of Montreal, whose book *Les Salaries Au Pouvoir!* explains their program in depth.

vii

The variety of communes, neighborhood corporations, and regional proposals now abundant within the United States are the first primitive attempts at self-determination. They are important for the questions they raise rather than the answers they provide.

But most important, they provide the political momentum which is absolutely required if ecological and social balance is to be achieved. They provide a working arena in which the issues of competition, scale, technology, and innovative government can be developed. In no sense can they be considered liberated zones. For social liberation cannot be achieved without biological liberation. That is to say that political autonomy depends upon the global political situation. There is no longer any way to separate the destiny of New York from that of any other large city in the world. These entities are intimately involved each with the other in an infinite number of exchanges, ranging from commodities and raw materials to culture and communication. Nor are the centers of urban development to be severed by any means from the vast areas of rural squalor. The metabolism of an Australian bushman is affected by mercury or strontium 90. The restoration of balance among people is inextricable from the restoration of balance between man and nature. The extent to which current experimentation begins to relate these parts of a single whole, begins to consider political and biological liberation as one and the same, is a measure of their promise for the future.

Liberation depends more than anything upon breaking the chain of domination at its strongest link here in the United States. There should be no illusion by any among the Left that the liberation movements of the developing countries will at some point cripple the United States. The new array of technological and chemical warfare agents being developed by the United States could under present conditions selectively devastate much of the Third World. At best, we in the United States must learn from the Vietnamese and other liberation movements the lessons of courage and dedication, and most important, the lesson of hope. In an age in which much is said about the overpowering effects of technology, the lesson of the Vietnamese people who have for a decade withstood the most intensive technological assault of all time is a truly miraculous example of the triumph of the collective human spirit against the will of the machine and those humans who are more machinelike in their actions that the machines they employ.

If it is true, as Heidegger and others have suggested, that technology is essentially nihilistic because it represents the ultimate expression of the will to power, it has been demonstrated in our own day as well that the collective will to defend a life of meaning has resisted the mechanical madness of American imperialism. Here perhaps is the most potent lesson for America today. The Vietnamese have withstood the pressure of death only because they resisted with life. By contrast, the gallows humor of almost any New Yorker reeks of the acceptance of a dead and decaying environment. This attitude of acceptance and resignation is manifold in the fear of the nuclear weapons or the inability to react to the biological destruction of the environment. The acceptance of death in the modern world stems from the ignorance of life. In other words, the failure for many in America to experience a truly human and creative environment, a life of meaning, robs a person of understanding the meaning of death as well.

What the struggle of the Indochinese people suggests is that only the struggle to defend life can counter the imminence of death. What this entails is the creation of a language and a program which can reach out and embrace the American public, which can serve as a source of gravity away from the banality of American life, away from the American way of death.

There must be no denying that the present moment and the one immediately ahead are full with awesome prospects. The French poet Pierre Emanuel at an American symposium perhaps captured the overwhelming contradiction, "America is prophet of the tension of the modern age and will have to suffer through it. America will help humanity to rediscover the meaning and necessity of suffering in a world of change." The prophetic tension of America is the contradiction between what exists and what is being born, between the cry for more survival on the part of frightened liberals, and the cry for life emanating out of the thousands of new life forms and experiments across America.

In the short run, the contradiction between death and life can only be heightened. The division of those who rule from those who are ruled is the first step toward liberation. That may well mean that in order to bring into being a new morning that a night of darkness more intense than anything yet endured will cover America. As long as the bombing of Indochina continues bombs will continue to explode in America. Certain critical installations of production, particularly chemical and petroleum sources, should be destroyed in the United States. It would be a grave miscalculation to believe that those who rule the American colossus will abandon their positions of power. The only way that the production of lethal agents will be terminated is if force is applied. In some cases short-term organizing strategies may disable key centers of production.

However, in many ecological industries, such as chemicals and petroleum, the highly capital intensive nature of production often rules out work place strikes as an effective tool. In such situations more severe forms of force should be applied. Key production sites should be destroyed through whatever means possible. There already have been selected incidents of such happenings at a petroleum refinery in New Jersey and explosions at a General Motors plant in California. Such activity is to be sharply distinguished from the less specific, more generalized wave of bombings that has taken place. What is being discussed here are not symbolic acts, but acts designed to terminate the operation of particular productive installations which are destroying the biosphere. It is a simple matter of self-defense.

The continued production of agents hazardous to the biosphere is worse than a game of Russian roulette, for we are certain that if it continues it will become terminal; we are only uncertain about the timetable. The sobering notations by scientists in Chapter Two should demonstrate that those who talk about a clean engine in the 1980's or later are in fact proposing a form of birth control on a scale unimagined by even Paul Ehrlich.

Such measures are stopgaps, nothing more. They do not contain within them the eggs of creation. The will to defend life, again, depends upon the existence of life. Beyond that dark night is a brilliant morning further in the distance. The path to that new morning was suggested by Rudi Dutschke, the German student leader, when he spoke about revolution in the industrial world requiring a "long march through the institutions."

Such a march must take the life force of America into virtually every institution, cutting across class and race, rural areas and urban areas. It will require what some have called a great refusal. But negation implies affirmation only barely. New forms of collective organization will be required to sustain and nourish the new society in the belly of the old.

In the 1960's a poet captured the prophetic tension of our age, "those not busy being born are busy dying." Today despair is the last resort of those who rule. Promise the first premise of those who struggle toward liberation. This is the promise of the great experiment briefly discussed in this chapter. It is a vision motivated by the remarkable abyss which separates the 1950's from the 1960's. We can see today in the children of the present moment a glimpse of the 1970's and the years ahead. Children for whom there is no going back.

In the backwaters of the continent's great rivers, the salmon annually make their way upstream, against tremendous currents, to spawn. They die, but life continues. The future of the present struggle cannot be seen at this moment, because we move toward such a beginning. It is our belief in the importance of that beginning which today confirms our inability to see the end.

Notes

CHAPTER ONE: The Power to Destroy and the Power to Create

The title for this chapter was taken from a statement of views by Ecology Action East, Box 344, Cooper Station, New York, N.Y. 10003. See also *Ecology and Revolutionary Thought* by Murray Bookchin, available from Anarchos, Box 466, Peter Stuyvesant Station, New York, N.Y. 10009.

[1]For a discussion of the politics of separation, see: Norman O. Brown, *Love's Body* (Random House, 1966), Ronald D. Laing, *The Politics of Experience* (Pantheon 1967), and an essay by Guy Debord, *The Society of the Spectacle* (France: the Situationists International).

[2]Loren Eiseley, *The Immense Journey* (Harcourt Brace and Jovanovich, Inc., 1957), p. 48.

[3]Maurice Merleau-Ponty, "Eye and Mind," *The Primacy of Perception,* ed. James M. Edie (Northwestern University Press, 1964), p. 187.

The philosophical and political writings of the French phenomenologist Maurice Merleau-Ponty influenced very much about this book. The above work, together with Maurice Merleau-Ponty, *Sense and Non-Sense,* trans. Hubert L. Dreyfus and Patricia Allen Dreyfus (Northwestern University Press, 1964), is a good introduction to Merleau-Ponty.

[4]"Special Edition on the Environment, *Fortune* (February 1970).

CHAPTER TWO: From Past to Present

[1]Ian McHarg, *Design With Nature* (Natural History Press, 1969), p. 24.

[2]John G. Neihardt (Flaming Rainbow), *Black Elk Speaks* (University of Nebraska Press, 1961), p. 1.

[3]The works of Lewis Mumford over a fifty-year period stand unrivaled as a definitive study of the origins and development of technological society. For our purposes, *Technics and Civilization* (Harcourt Brace Jovanovich, 1934), *The City in History* (Harcourt Brace Jovanovich, Inc., 1961), and the *The Myth of the Machine: Pentagon of Power* (Harcourt Brace Jovanovich, Inc., 1970) are of particular interest.

[4]Alexis de Tocqueville, *Democracy in America,* vol. 2 (Vintage Books, 1954), pp. 336-337.

[5]Perry Miller, *Errand Into the Wilderness* (Harper Torchbooks, 1964), p. 239.

[6]Samuel P. Hays, *Conservation and the Gospel of Efficiency* (Harvard University Press, 1959), p. 265.

[7]Gabriel Kolko, *The Triumph of Conservatism* (The Free Press, 1963), p. 3.

[8]James O'Connor, "The Fiscal Crisis of the State," *Socialist Revolution*, vol. 1, no. 1, p. 20.

[9]Hays, *Conservation*, p. 271.

[10]Barry Weisberg, "A One Day Teach-In Is Like An All Day Sucker," *Liberation,* (May 1970).

[11]U.S. Department of Health, Education and Welfare, *Community Water Supply,* Significance of Major Findings, Bureau of Water Hygiene (Washington, D.C., July 1970).

CHAPTER THREE: The Structure of Social and Ecological Responsibility

[1]Ernst Cassirer, *The Question of Jean Jacques Rousseau,* trans. Peter Gay (Indiana University Press, 1963).

[2]E. J. Mishan, *The Cost of Economic Growth* (Frederick A. Praeger, 1967), and Paul Baran, *The Political Economy of Growth* (Monthly Review Press, 1957).

[3]René Dubos, *Man, Medicine, and Environment* (Frederick A. Praeger, 1968), p. 7. For a more generalized overview of man's place in the environment read Rene Dubos, *So Human an Animal* (Charles Scribner's Sons, 1968). For those concerned with issues of mutual aid, Peter Kropotkin's classic *Mutual Aid* is the basic text.

[4]Andre Gorz, *Strategy for Labor* (Beacon Press, 1967), p. 95.

[5]Ibid., p. 68.

[6]James O'Connor, "The Fiscal Crisis of the State," *Socialist Revolution,* vol 1, no. 1, p. 19.

[7]Fred S. Singer, *Global Effects of Environmental Pollution: A Symposium Organized by the American Association for the Advancement of Science, December, 1968* (New York: Springer-Verlag, 1970), p. 4. From a paper by Francis S. Johnson, "The Oxygen and Carbon Dioxide Balance in the Earth's Atmosphere." This is the best scientific overview of current global ecological conditions.

[8]Ibid., p. 22, from a paper by F. D. Sisler, "Impact of Land and Sea Pollution on the Chemical Stability of the Atmosphere."

[9]Ibid., p. 179, from a paper by Edward D. Goldberg, "The Chemical Invasion of the Oceans by Man."

[10]Ibid., p. 4, Johnson. See note 7.

[11]Ibid., p. 188, from a paper by George M. Woodwell, "Changes in the Chemistry of the Oceans: The Pattern of Effects."

[12]Ibid., p. 179, Goldberg. See note 9.

[13]Robert Heilbroner, "Marxism: For and Against," *New York Review of Books* (June 5, 1969), p. 16.

[14]Paul Baran and Paul Sweezy, *Monopoly Capitalism* (Monthly Review Press, 1966), p. 389.

[15]Gorz, *Strategy for Labor,* p. 91. See note 4.

CHAPTER FOUR: The War Machine

[1]Glimpses of how the Department of Defense and the Pentagon operate on American society are offered by Richard J. Barnet, *The Economy of Death* (Atheneum, 1969); Seymour Melman, *Pentagon Capitalism* (McGraw Hill, 1970); and William Proxmire, *Report From Wasteland* (Praeger, 1970).

[2]Why there is no military-industrial complex is offered in a special report of the Bay Area Institute. Martin Gellen, *The Permanent War Economy* (San Francisco, 1971). Unpublished manuscript.

[3]Richard J. Barnet, *The Economy of Death* (Atheneum, 1969), p. 42.

[4]Paul Jacobs, "Precautions Are Being Taken by Those Who Know," *The Atlantic* (February, 1971), p. 56.

[5]Mumford, p. 96.

[6]Barry Weisberg, ed., *Ecocide in Indochina: The Ecology of War* (Canfield Press of Harper and Row, 1970).

[7]Thomas Whiteside, "Defoliation," *The New Yorker* (July 11, 1970).
[8]Harry Magdoff, *The Age of Imperialism* (Monthly Review Press, 1969), pp. 185-186.
[9]Martin Gellen, *The Permanent War Economy.*

CHAPTER FIVE: Oiling the Machine: Automobiles and Petroleum

[1]The political economy of petroleum — while the largest segment of the world economy — is comparatively unexamined. Introductory texts would include: *The Politics of Oil,* Robert Engler (University of Chicago Press, 1967); *The Political Economy of Oil in Underdeveloped Countries,* Michael Tanzer (Beacon Press, 1969); *The Empire of Oil,* Harvey O'Connor (Monthly Review Press, 1970); and *Le Petrole, Nouvel Enjeu de la Decolonisation,* (Le Monde Diplomatique, March 1971).
[2]Emma Rothschild, "GM in Trouble," part one, *New York Review of Books* (February 25, 1971), p. 17. Part two will appear in a future edition. Other introductions to the automotive industry and General Motors include *1970 Automobile Facts and Figures,* published by the Automobile Manufacturers Association, Detroit; *Giant Enterprise, Ford, and General Motors, and the Automobile Industry,* Alfred D. Chandler, Jr. (Harcourt Brace Jovanovich, 1964); *My Years With General Motors,* Alfred B. Sloan (Doubleday & Co., Inc., 1964); *What's Good for GM,* Edward Ayres (Aurora Press, 1970).
[3]John McHale, "World Energy Resources in the Future," *Futures,* vol 1, no. 1 (September 1968), p. 11.
[4]"The Fantastic Rise in Corporate Profit, *AFL-CIO Federationist* (May 1966), a pamphlet.
[5]McHale, "World Energy Resources," p. 5.
[6]Barry Weisberg, "The Offshore Oil of Southeast Asia," *Nation* (March 8, 1971). See also Malcolm Caldwell, "Oil and the War," *Liberation* (Spring 1971).

CHAPTER SIX: The Conditions of Liberation

[1]Warren Winkelstein, Jr., and Fern E. French, "The Role of Ecology in the Design of a Health Care System," *California Medicine* (November 1970), 113:7. For further investigation of this perspective, examine the workings and writings of the Health Policy Advisory Center (Health-PAC), 17 Murray Street, New York, N.Y.
[2]John Gurley, "Capitalist and Maoist Economic Development," *Bulletin of the Concerned Asian Scholars* (April-July 1970), vol. 2, no. 3, p. 34. This article also appears in *America's Asia,* edited by Edward Friedman and Mark Selden (Pantheon, 1971). The Gurley article contains numerous references for further examination of the issued raised by my reference.
[3]Ibid., p. 47.
[4]Writing Group of the Tientsin Municipal Revolutionary Committee, "Multiple Use: Important Policy for Industrial Production," *Peking Review,* 6 (February 5, 1971), p. 7.
[5]Ibid., pp. 9-10.
[6]Shen Wen, "Going in for Farming in an Industrial City," *Peking Review,* 10 (March 5, 1971), pp. 8-9.

[7]See Robert Hinton, *Fanshen: A Documentary of Revolution in a Chinese Village* (Vintage, 1968).

[8]Anyone interested in workers' control should begin with *Strategy for Labor* by Andre Gorz (Beacon Press, 1964). Gorz has a recent article on workers' control in America "Workers' Control," *Socialist Revolution*, vol. 1, no. 6, November-December 1970.

[9]Two descriptions of this process are offered in Gerard Chaliand, *The Peasants of North Vietnam* (Penguin, 1969), and Jon M. Van Dyke, *North Vietnam's Strategy for Survival* (Pacific Books, 1971).

[10]An introduction to the socialist economy of North Korea is offered by the Head of State Planning in the Democratic People's Republic of Korea, Yun Gi Bok, "Socialist Economy in Korea," *Tricontinental*, No. 19-20.

[11]Kim Byong Sik, *Modern Korea: The Socialist North, Revolutionary Perspectives in the South and Unification* (International Publishers, 1970). Sik explains the principle of juche, "Kim Il Sung's revolutionary ideas are based on the concept of Juche. . . . This concept embraces the principle of self-reliance in ideology and politics, in the economy and in military affairs. It means to think creatively about all problems ourselves, to solve them in accord with actual conditions in our country, independently with our own strength, and in accord with our own revolutionary interests," pp. 10-11.

[12]The Chongsanri Method and Spirit and the Adean system of socialist management are explained in *Modern Korea*, pp. 113-131.

[13]A provocative critique of Marxist thought from a post-scarcity perspective is "Listen Marxist!" a pamphlet available from Anarchos, Box 466, Peter Stuyvesant Station, New York, N.Y., 10009.

[14]Murray Bookchin, *Post-Scarcity Anarchy* (Ramparts Press, 1971).

[15]Herbert Marcuse, "Political Preface 1966," *Eros and Civilization* (Beacon Press, 1966), p. xvi.

[16]Murray Bookchin, "Toward a Liberatory Technology," a pamphlet available from Anarchos, see note 13.

[17]Herbert Marcuse, "The End of Utopia," *Five Lectures* (Beacon, 1970), pp. 63, 78.

[18]Introduction, *Socialist Revolution*, vol. 1, no. 6, p. 7.

[19]Ibid., p. 7.

[20]Ray Davidson, *Peril on the Job* (Public Affairs Press, 1970), p. 1.

[21]Introductions to the League of Black Revolutionary Workers include "Black Labor," an edition of *Radical America*, vol. 5, no. 2, March 1971, and "Our Thing is Drum!" *Leviathan*, vol. 2, no. 2, June 1970; also available as a pamphlet by the same title from Black Star Publishing, Detroit.

[22]Herbert Marcuse, *An Essay on Liberation* (Beacon, 1970), pp. 21-22.

[23]Milton Kotler, *Neighborhood Government, The Local Foundations of Political Life* (Bobbs-Merrill, 1969).

[24]Ibid., p. 39.

[25]Ibid., p. 76.

[26]Peter Manso, editor, *Running Against the Machine* (Doubleday and Co., 1969), p. 6.

Index

The Accidental President: 131
Acheson, Dean: 137
Adams, John: 20, 24, 42
Advertising: 55; dominance of American, 62
Affluence, model of: 74
Africa, wildlife in: 31
Agence France Press: 142
Air pollution: morbidity and, 117; mortality and, 117
Air pollution control, funds for (table): 34
Alaska: 132, oil development in, 137
Alaskan pipeline, public struggles against: 99. See also TAPS Alaskan pipeline
Alka-Seltzer: 55
Amchitka Island, nuclear testing on: 93
America: ecological policy of, 56; as empire, 57, 148; foreign policy, 98; international economy of, 98; protest in, 168; resignation in, 181; social disintegration of, 165; social diversity in, 169; utopian communities in, 169; Vietnamese resistance to, 181. *See also* United States
American Association for the Advancement of Science (AAAS): 65-65, 90
American Motors Corporation, antitrust suit against: 36
American Transit Association: 119
Anderson, Jack: 138
Anderson, Robert O.: 133-34, 136
Antitrust Subcommittee, Senate: 121
ARCO. *See* Atlantic Richfield
Armed Forces: 81
Asbestos: and industrial diseases, 164; lung disease and, 117
Asia, consumption of: 65
Aspen Institute: 134
Assembly line: 78, 100; hazardous conditions on, 164
AT&T (American Telephone and Telegraph): 59, 94, 107
Atlantic Oil Storage Company. *See* Atlantic Richfield (ARCO)
Atlantic Refining Company. *See* Atlantic Richfield (ARCO)
Atlantic Richfield (ARCO): 131, 132-34, 136, 137, 143; foreign markets of, 133; and no-lead gasoline, 133; political campaign contributions of, 134; primary properties of, 133; and Robert O. Anderson, 133-34
Atmosphere, injection of materials into: 69
Atomic bomb, development of: 77
Atomic Energy Commission (AEC): 83-85, 88; employees of, 88
Attlee, Clement: 85
Australian aborigine, philosophy of: 15
Authority: 50-51
Automobiles: advertising of, 103-4; alternate propulsion systems for, 115-16; and consumer concern, 104; cultural effects of, 108; dependence upon, 113; efficiency of, 104; engineering innovations in, 104; exhausts, 9; influence of, 62; limiting, 176-77; political economy of, 2; pollution control for, 115-16; pollution free engine for, 115; as pollution source, 64; recall of, 104; safety devices for, 114-15; social costs of, 117; social effects of, 109; taxes on, 114; use of, 74; as waste, 49. *See also* Motor vehicles
Automobile industry: 10, 58, 76; employment in, 103; growth of, 101
Automobile Manufacturers Association: 101; antitrust suit against, 36
Automotive industry: 99, 112; and American policy, 136
Autonomy. *See* Self-determination
Aviation industry: 78

Bacon, Francis: 17
Balance: 181; biological, 47; of ecology, 10; social, 47
Baran, Paul: 49, 73

Barnet, Richard J.: 80
Barriers, social: 109
"Benign neglect," policy of: 33
Berkeley, California: 165-66
Bible, Senator Allan: 127
Bible bill: 127
Big Four: 119. *See also* General Motors Company, Ford Motor Company, Chrysler Corporation, American Motors
Big Oil: 138
Bikini Atoll, atomic testing at: 83
Biosphere, potential life of: 7
Bismarck, Otto von: 76
Black Elk: 53
"Black power" (in oil industry): 131
Blue Cross: 71
Boeing Aircraft Corporation: 62, 92, 94, 96, 119
Bolsheviks, opposition to: 79
Bombings: 183
Bookchin, Murray: 165
Boone, Daniel: 20
Bottles, as waste: 49
Bourgeois revolution: 16
Bread and circuses: 171
Breslin, Jimmy: 180
Brest Litovsk, Treaty of: 79
Britain: 59, 147
British Petroleum: 138
"Broken Arrows": 83
Brower, David: 134
Brown University: 124
Buber, Martin: 160
Burke, William O.: 58
Business Week: 43, 59, 82, 129

C-5A aircraft: 81
C-123 cargo planes: 91
Calhoun, J. B.: 146
California, Southern: 30
California Standard. *See* Standard Oil of California
Calorie intake: 44
Cambodia: 90; invasion of, 11, 30, 141
Canada: 31, 138
Cancer: 84, death rates from, 9
Capital: 61; accumulation of, 71; circulation of, 73; distribution of, 44; and resource distribution, 120; state and corporate, 72
Capitalism, development of, 53-54; division of labor under, 162-63; and ecological imbalance, 149, 160; ecology of, 145; and freedom, 164-65; political, 112; societal organization and, 150; war and, 78
Carbon dioxide, and industrial diseases, 164; as pollutant, 64
Carbon monoxide: emissions of, 63, 69; levels of, 115; as pollutant, 64; tolerance for, 89
Carcinogens: 9
Cartesian method: 17
Case Western Reserve University: 124
Castlemanship: 19
CBW. *See* Weapons, biological; Weapons, chemical
Cement: 110
Central Intelligence Agency (CIA): 83, 137; in Middle East, 139
Centralization: 170
Central Treaty Organization (CENTO): 139
Chase Manhattan Bank: 111, 134, 143
Chemicals, production of: 38
Chicago Institute of Medicine: 163
China, Communist: 149; conservation of resources in, 153; economic development in, 150; industry and agriculture in, 153-54; "multiple use" in, 151-52; social organization in, 179; and "socialist man," 151; trade with, 58; waste in, 152; worker-peasant alliance in, 154
Chinese Cultural Revolution: 154
Chrysler Corporation: 106; antitrust suit against, 36
Churchill, Winston: 85
Cigarettes, consumption of: 71
Cities: Federal colonization of, 173; inner, 51; large, 61; problems of, 30
Cities Service: 143
Citizens Board of Inquiry on Hunger and Malnutrition: 43

Classes: formation of, 1; hierarchy of, 54
Classless society: 154, 161
Class struggle: 162
Clean Car Race: 116
Coal: 121-22; industrial nations' consumption of, 138; production of, 123
Coca Cola: 62
Coexistence: 53
Cold War: 77
Collective organization: 183
Columbia University: 124
Columbus, Christopher: 18
Commodities, fetishism of: 55
Common Market nations, oil consumption of: 139
Communautés des travails: 156
Communes: 169
Communication, military requirements for: 78
Communism, and human development: 159-60
Communist man: 154
Communist Manifesto: 161
Community: 169; biological imbalance of, 171
Community control: 176
Community Water Supply: 37
Competition: 53
Congressional Record: 124
Connor, John T.: 112
Conservation: 23; concepts of, 28-29; organizations for, 30-32
Conservation and the Gospel of Efficiency: 23
Constitution, early: 19-20
Consumer power: 70, 75
Consumers: 54; concerns of, 10-11, 104
Consumer sovereignty, myth of: 49
Consumption: American, 70; military, 77; society and, 74; society of, 52, 54
Contaminants, atmospheric: 70
Contamination: air, 64; among Alaskan seals, 56
Continental Coal: 121
Contractors, highway and construction: 113
Conversion: economic, 96; proposals for, 95
Copernican revolution: 17
Corporate state: 76
Corporations: giant, 73; and government, 112
Cosmetics; advertising of: 55
Council on Economic Advisors: 73
Council on Environmental Quality: 35-36
Credit, consumer: 100
A Critique of Pure Tolerance: 21
Crockett, Davy: 20
Cuba, U.S. invasion of: 137
Culture, and automotive-petroleum empires: 99

Dartmouth University: 124
Datsun: 117
Davidson, Ray: 164
DDT: 1, 72; in Antarctic, 56; in human milk, 3; in pelican eggs, 66; sale of, 67
Death in Life: 83
Decentralization: 120, 170; in China, 150-52, 157; in Cuba, 157; in North Korea, 157; in North Vietnam, 157
Decisions: 4
Decornoy, Jacques: 143
Defoliants: government inspection of, 90; in Vietnam, 89
Democracy, despotism of: 20-21, 109
Democracy in America: 20
Democratic Party: 92
Democratic People's Republic of North Korea (DPRK). *See* North Korea
Depletion allowance: 129-30
Descartes, René: 17
Design with Nature: 171
Destruction: 53
Detergents: 110
de Tocqueville, Alexis: 20-21, 109
Detroit, carbon monoxide in: 63
Developing countries: liberation movements in, 181; social organization in, 179
Dialectical materialism, social and

ecological implications of: 151
Dillingham Environmental Company: 67
Distane Early Warning (DEW) Line, ecological implications of: 82
Distribution, of commodities: 78
Diversification: 170; in China, 154
Division of Labor: 74, 162-63
Djakarta: 135
DNA: 86
Donora, Pa., deaths from pollutants in: 64
The Double Helix: 86
Doxiadis, C. A.: 108
Dresden 1 Nuclear Station, in Chicago: 84
Drugs, advertising of: 55
Dubos, René: 53
DuBridge, Lee: 33
Duchamp, Marcel: 2
Dugway proving ground: 88
Dulles, Allan: 137
Dulles, John Foster: 137
Dutschke, Rudi: 183

Earth Day: 31
East Central Citizens Organization (ECCO): 170-71
Eastern Airlines: 111
Ecocide, in Indochina: 90
Ecocide in Indochina: The Ecology of War: 89
Ecological balance: 10, 160
"Ecological crisis": 41
Ecological imbalance, and oil industry: 136
Ecological industries, destruction of: 182-83
Ecological responsibility: 44, 70; class structure of, 43-45
Ecology: disorder of, 147; human destiny and, 11; imbalance of, 49, 57
Economics, Keynesian: 94
Economist: 73
Economy: atomobile manufacture and, 100; and automotive-petroleum empires, 99
The Economy of Death: 80
Ecosystem: 2; exploitation of, 3; tolerance level of, 6
Edgewood Arsenal: 88
Edwards, Jonathan: 22
Efficiency, and American economy: 60
Ehrlich, Paul: 183
Eichmann, Adolf: 42
Einstein, Albert: 86
Eisely, Loren: 3
Eisenhower, Dwight David: 130, 137, 141; administration of, 45
Electrical World Week: 127
Electric power consumption: 43
Electronics industry: 78
Emerson, Ralph Waldo: 20-22
Emmanuel, Pierre: 182
Emphysema: 9, 64
Employment, on mass assembly line: 105
"The End of Utopia": 162
Energy: costs of, 122; income sources of, 125-26; industrial supply of, 138; safe form of, 124
Energy consumption: per capita rate, 43; ratio of, 69
Energy industry, university holdings in (table): 124
Eniwetok, atomic testing at: 83
Environment: Federal expenditures on, 33-34; Federal programs and, 40; man's domination of, 20; ownership of, 73
"Environmental crisis": 5
Environmental Policy Act, of 1969: 35
Environmental Quality Council: 115
Equitable Life Insurance: 111
Errand into the Wilderness: 22
Esso: 59, 139, 144
Ethyl Export Corporation: 79
Europe, consumption of: 65
Evans, Brigadier General William John: 142
"Externalities": 72

"Fail safe": 88
Fallout, atomic: 57
Family planning: 48
Fanshen: 156
Farmers, price support: 44
Federal programs: 33-34; environmental, 38
Federal Radiation Council: 85
Federal Water Quality Administration: 35

Fertilizers: 110; nitrogen, 67
Fetus, deforming of: 89
First National Bank of Boston: 107
First National Bank of St. Louis: 107
First National City Bank of New York: 107, 112
Fisheries, oceanic: 67
"Flywheel energy storage system": 116
Food, unequal consumption of: 70
Food Conspiracy, the: 166
Food and Drug Administration: 55-56
Forbes magazine: 35
Force, use of: 51, 182-83
Ford, Henry: 100
Ford, Henry II: 58
Ford Asia-Pacific-South Africa, Inc.: 58
Ford Motor Company: 59, 58, 94, 100, 106, 136; antitrust suit against, 36
Fort Dietrick: 88
Fort McClellan, Alabama: 87-88
Fortune magazine: 4, 41, 60, 98, 107
Fossil fuel, dependence on: 125
Fourier, Charles : 161, 165
France: 59, 147, 178; May-June 1968, 165; workers' control in, 156
Freedom, precondition of: 164
Freeman, David: 123
Freon engine: 117, 127
Friends of the Earth: 30, 41, 134
Le Front D'Action Politique: 180
Frontier, conquering of: 20
Fuel oils: 110
Fulbright, Senator J. William: 71
Fuller, Buckminster: 61
Fungicides: 37

Gas, natural. *See* Natural gas
Gasoline, Federal tax on: 113
Gellen, Martin: 80, 94, 96
General Motors (GM): 53, 59, 76, 79, 94, 100, 106, 132, 136; Acceptance Division of, 107; advertising of, 103; antitrust suit against, 36; nonautomotive production of, 106; profits of, 108; strikes against, 107
Genesis, story of Creation: 13-14
Genetic engineering: 86
Geological Survey, U.S.: 126
Geothermal energy: 126-28; UN Conference on, 127
Global Effects of Environmental Pollution: 65
Gofman, John: 84-85
Gorz, Andre: 54, 74, 167
Government: and corporations, 112; Federal, 172-73; and ecosystem, 171; neighborhood, 171-72; purpose of, 19; regulation of, 119; state, 172-73
Government-employee unions, strikes in: 173-74
Government Operations, House Committee on: 35
Greeks, ancient: 13
Gross National Product (GNP): 10, 60, 99, 150; automobile industry and, 102; in 1830, 22; in 1880,22; increase in, 60, 71; militarism in, 94; surplus and, 73; of U.S., 59
Growth: 73, 161
Guatemala: 137
Gulf of Thailand, oil exploration in: 142-43
Gulf Oil: 107, 124, 136, 138-39
Gurley, John: 149-51, 154

Hamilton, Alexander: 57
Hays, Samuel P.: 23, 25, 29
Health: 71
Hearst Press, the: 31
Heidegger, Martin: 181
Heilbroner, Robert: 71-72
Heraclitus: 13, 24
Herbicides: 110; forced urbanization and, 91; Senate vote concerning, 91; in South Vietnam, 89
Herter, Christian A.: 137
Hetch Hetchy Valley Yosemite: 24
Hickel, Walter: 137
Highways: accidents on, 114; deaths on, 8; expenditure for, 112-13
Highway system: social implications
Highway Trust Fund: 112-13, 119, 128
Hilton hotels: 62
Hiroshima: 1
"Hiroshima Generation": 83
History: Marxist view of, 161; in schools, 53; stages of, 46

Holifield, Chet: 85
Hotels, international: 62
House Ways and Means Committee: 39
Household, space in: 177-78
Hudson River, pollution of: 36
Humble Oil: 121
Hutchinson, G. Evelyn: 7
Hydrocarbons: 69, 100; lung disease and, 117
Hydroelectricity, industrial nations' consumption of: 138
Hydrological energy: 125

Iceland, hot-water energy in: 127
Ickes, Harold Le C.: 79
I. G. Farben: 79
Imbalance: ecological, 97-98; within human society, 147; within natural systems, 147; social, 97
"Import tickets": 130. *See also* Oil industry, import controls in
Income, personal: 103
Indians, American: 18
Individualism, capitalist ethic of: 108
Indochina: American economic investment in, 142; American oil interests in, 99; American petroleum exploration in, 141; ecological warfare in, 89; liberation of, 11; oil deposits in, 133; and petroleum, 138; war in, 9, 81, 93. *See also* South Vietnam, Cambodia, Laos
Indonesia, U.S. oil interests in: 143
Industrial corporations: dollar comparisons of (table), 107; largest, 107
Industrial Revolution: 16
Industry: chemical emissions of, 35; consumption of materials in automobile, 101; deaths in, 164; defense sector and, 94; diseases in, 164; employment in automobile 103; energy, 121; glass, 102; growth of automobile, 101; hazardous conditions in, 163-64; human isolation and, 78; military perspective of, 96; sewage emissions of, 45; toxic chemicals in, 164; war, 76; wastes generated by, 4-5. water required for, 70; workers' unrest in, 164
Infant mortality, nuclear wastes and: 84
Inflation: 119, motor vehicle industry and, 103
Institutions: 42-43
Intercellular altruism: 24
Interdependence: 2, 53
Interior and Insular Affairs, Senate Committee on: 92
Internal combustion engine, dependence on: 121
International Business Machines (IBM): 94, 107
International corporations: 58
Interstate Commerce Commission: 36
Interstate highway system: 113
Investment, foreign: 58
Iran: 137
Irwin, John: 137
Israel, workers' control in: 156
"The Issue That Could Unite the Generations": 31

Jackson, Senator Henry (Scoop): 32, 92-93
Jakarta: 135
Japan: 178; consumption of, 66; oil consumption of, 138; petroleum industry and, 79; and U.S. oil interests, 144
Jaspers, Karl: 7
Jersey Standard. *See* Standard Oil of New Jersey
Jet set: 62
John Muir Institute: 134
Johns Hopkins University: 116
Johnson, Francis S.: 64-66
Johnson, Lyndon B.: 127, 137; administration of, 5, 45
Juche, North Korean concept of: 158
Judeo-Christian era, advent of: 13
Justice Department, economic concentration and: 27

Das Kapital: 55
Kastenmeier, Robert W.: 87
Kennedy, John F.: 42, 92, 137; administration of, 45

Keyes, Roger: 112
Keynes, John Maynard: 150
Kim Il Sung: 158-59
Kissinger, Henry: 137
Kolko, Gabriel: 26
Kotler, Milton: 170, 172

Labor: division of, 74, 162-63; social division of, 1
Laing, R. D.: 50, 168
Landlord, slum: 9
Laos: bombing of, 89; invasion of, 142. *See also* Indochina
Lapp, Ralph: 83
Lave, Lester B.: 117
Law, force and: 20
Lead: anti-knock, 67; in gasoline, 72; in surface waters, 68
League of Revolutionary Black Workers: 164
Lear, John: 127
Lehigh University: 124
Lenin, V. I., and ecology: 149
Leukemia: 84
Levy, Walter J.: 140
Liberation: biological, 170, 181; and force, 182-83; political, 48, 181; precondition for, 32; social, 170
Liberty Against Organization: 76
Libya: 139
Life: alienation of, 108; matter and, 3
Lifton, Robert Jay: 83
Liquor, consumption of: 71
Listerine: 55
Litter, recycling: 49-50
Litton Industries: 136
Livermore Laboratory, of Atomic Energy Commission: 84
Lockheed Aircraft Corporation: 94
Lodge, Henry Cabot: 141
London, deaths from pollutants in: 64
Lon Nol, General: 142
Lubricants: 110

Machiasport: 132
Madison, James: 19, 28, 42
Madisonian interest politics: 150
Magdoff, Harry: 94
Mailer, Norman: 173, 180
Male chauvinism: 10
Malaysia, oil discoveries in: 143-44
Malnutrition: 1, 43-44
Mandel, Ernst: 71
Manhattan: 120
Manhattan oil tanker venture: 133
Mann, Thomas: 112
Maoism, ecological implications of: 155
Mao Tse-tung: 151
Marcuse, Herbert: 21, 52, 161, 165
Market: control of, 60; Federal regulation of, 26; petroleum, 110; military, 81
Marshall Islanders, thyroid disorders of: 85
Marx, Karl: 55, 73, 147, 150, 161-62, 165
Marxism: historical view of, 161; necessity in, 161; social hierarchy in, 162
The Masculine Birth of Time: 17
Massachusetts Institute of Technology (MIT): 124
Mass production: 77
Mass transit, diversified system of: 118, 176
Mass transit systems: 119
Masters of Capital: 111
McGraw-Hill Publications: 152-53
McHarg, Ian: 14, 171
McNamara, Robert: 112, 136
Mechanics: 102
Medicine, American: 9
Megamachine: 16
Mekong River Delta, petroleum deposits in: 141
Mellon, Richard King: 136
Mellon National Bank and Trust Company: 107
Melville, Herman: 20
Mercury: 56
Mercury contamination: 67
Merleau-Ponty, Maurice: 3
Metcalf, Senator Lee: 121, 124
Metroliner: 120
Metropolitan Life Insurance: 111
Meuse Valley, Belgium, deaths from pollutants in: 64
Mexicans: 18

Mexico City, carbon monoxide in: 63
Middle class: American, 29; food discarded by, 44-45
Middle East: 123, 129, 133, 138; American military presence in, 139; American oil interests in, 99; national petroleum companies in, 140; protecting U.S. oil interests in, 138
Miles Laboratories: 55
Militarism, costs of: 89
Military: capital investment and, 82; ecological policy and, 92; in industrial society, 77; spending of, 78, 94; transportation and, 118
Military bases: costs of, 81; effects of, 82; extensive system of, 82
Military Construction Authorization Act: 93
Military-industrial complex: 80, 95
Miller, Perry: 22
Mills, C. Wright: 128
Minorities, plight of: 22
Minto, Wallace L.: 116-17
Mishan, Ezra: 49
Mitchell, Attorney General John J.: 59, 131, 138
Mobil Oil: 107, 111, 124, 138-39
Le Monde: 143
Monopoly: 26, 57, 76; energy, 122; of oil industry, 123
Monopoly Capital: 73
Montreal Trust Company: 108
Moody, John: 111
Moore, Barrington, Jr.: 21
Morgan Guaranty Trust Company: 107
Moscow: 149
Motor vehicles: consumption of, 105-6; production of, 100, 102, 105. *See also* Automobiles
Moynihan, Daniel Patrick: 33
Muir, John: 22, 24, 29
Multiple-purpose appliances: 177
"Multiple-use": 151-52, 157
Mumford, Lewis: 16, 18, 77-78, 171
Mussolini, Benito: 79
Myrdal, Gunnar: 80

Nader, Ralph: 71
Nagasaki: 1
National Bank of Detroit: 108
National defense: 5; ecological and social implications of, 80-82 spending for, 80
National Economic Research Association: 121
National Industrial Pollution Control Council: 36
National Institutes of Health: 89
National Organization of State Highway Officials: 113
Natomas of California: 143
Natural gas: 121; industrial nations' consumption of, 138; U.S. consumption of, 138
Natural resource programs, Federal budget for (table): 33
Nature: Eastern view of, 155; political management of, 8; Puritans and, 18; subjugation of, 14
Nature conservancy: 30
Neighborhood: 169-72
Neighborhood Government: 170
Nelson, Senator Gaylord: 116
Nerve gas: 87
Nevada, atomic testing in: 83
The New Atlantis: 17
"New Jerusalem": 18
New World, exploration of: 15-16
New York City: 146-47; deaths from pollutants in, 64; expenditures of, 173; as state, 180; taxes of, 173
New Yorker: 67
New York Times: 45, 111, 169
New Zealand, geothermal sources in: 127
Newsweek: 148
Nitrogen oxide: 66
Nixon, Richard: 33, 36, 113, 127, 131, 134, 136-37, 141; administration of, 35
North Korea: Chonsanri Spirit in, 159; Adean System in, 159; self-reliance in, 158-59
North Slope, Alaska: 131, 133
Northwestern University: 124
Nourse: Edwin G.: 73
Nuclear accidents: 83
Nuclear power, nonpolluting energy from: 123

Nuclear preparedness, theory of: 92
Nuclear reactors: 85
Nuremburg War Crimes Trials: 79

Oblivion, rush to: 80
Obsolescence, built-in: 56
Oceanic system: degradation of, 67
Oceans: 3; lead in, 67-68; as nuclear dumps, 83, 88
O'Connor, James: 28, 39
Oil: crude, 104; and Indochina war policy, 142-43; shortages, 122. *See also* Oil industry, Petroleum, Petroleum industry
Oil, Chemical, and Atomic Workers: 31
Oil and Gas Journal: 131
Oil capitals: 135
Oil corporations, net income of: 98
Oil industry: 110; as American foreign investment, 135; depletion allowance in, 129-30; and ecological imbalance, 136; and economic colonization, 135-36; imports controls in, 129-30; prorations in, 129-30; social domination of, 128; and underdeveloped countries, 135-36; and U.S. State Department, 137-38. *See also* Petroleum industry
Oil spills: 99
Okinawa: 82
Oligopoly: 26
On Contradiction: 151
Opel: 106
Operation Ranchland: 90
Organization, social: 2
Organization of Petroleum Exporting Countries (OPEC): 137-39
Orians, Gordon: 90
Original sin, conception of: 42

Pacific Gas and Electric: 127
Paints: 110
Paper, as waste: 49
Parmenides: 13
Patriarchy: 15
Pax Americana: 62
Peking Review: 152
Pelican: 66
Pentagon: 78; employment turnover of, 81; selling of, 82
The Pentagon of Power: 16
People's Architects: 166
People's Park: 165-66
Perfumes: 110
Perils on the Job: 164
Persian Gulf settlements: 140
Persian Gulf states: 137
Pertamina: 143
Pesticides: 37; public struggles against, 99; U.S. production of, 56
Petrochemicals: 110
Petroleum: Asian consumption of, 138; industrial nations' consumption of, 138; political economy of, 2; and production and consumption, 109; reserves of, 131; supply and demand of, 122; U.S. consumption of, 138; uses of, 109. *See also* Oil, Oil industry, Petroleum industry
Petroleum industry: 10, 58, 76, 99, 111-12, 121; and American policy, 136; Federal incentives for, 128-30; foreign investment by, 139; and Indochina, 132; political campaign contributions of, 136; sales of, 111. *See also* Oil, Oil industry, Petroleum
Pfeiffer, E. W.: 90
Pharmaceuticals: 110
Philippines, military installations in: 82
Philosophy: holistic, 32; Oriental, 15
Photosynthesis: 125
Pinchot, Gifford: 22-26
Pine Bluff Arsenal: 88
Poland, oil deposits in: 79
Polarization: 32
The Political Economy of International Oil and the Underdeveloped Countries: 135
Politics: 8, 171
The Politics of Experience: 50
Pollutants, deaths from: 64
Pollution: concept of, 5; fines for, 34; public struggles against, 99; rates of atmospheric (table), 66; thermal, 85; in union contracts, 180

Pollution control: 5, 30, 46, 75
Population bomb: 11, 41
Population control: 47-48, 115
Population Control Through Nuclear Population: 84
Population growth: 47
"Power crisis": 122
Power nexus, elimination of: 167
Power plants: 85
Power shortages: 123
Princeton University: 124
Product, added value of: 74
Production: alternate forms of, 177; geared to death, 178; limits on, 175-76; social organization of, 61; standarization of, 77; techniques of, 100
Profit: 61, 119
Progress, notion of: 16-17
Progressive Era: 22-23, 26-28, 30, 36, 41, 72
Proletarian communism: 159
Proletariat: 163
Protest: 10, 168
Proxmire, Senator William: 81, 95
Prudhoe Bay. *See* North Slope, Alaska
Public service systems: 40
Puritans: 18

Rabenhorst, David W.: 116
Radiation: effects of 84; Federal standards for, 84; reduction of, 86.
Railroads: 100, 118
Rankine cycle engines: 116. *See also* Freon engine
Rationing, of resources: 174-75
Reader's Digest: 167
Reality of Constant Threat: 11
Rebellion: 52
Reconversion, proposals for: 95
Redistribution, of wealth and resources: 47
Refuse Act of 1899: 35
Regional organizing: 180
Regulation, Federal: 75
Report from Wasteland: 81
Resources for the Future: 134
Revolution: 11; U.S. campaign against, 83
Revolutionary consciousness: 162
Ribicoff, Senator Abraham: 96
Richfield Oil. *See* Atlantic Richfield (ARCO)
Ritual: 50
River Thames: 6
Rockefeller Council on Foreign Relations: 137
Rockefeller family: David, 143; economic empire of, 111; John D.: 76, 110; Nelson: 137
Rockefeller Foundation: 137-38
Rockefeller interests: and American policy, 137; economic empire of, 111
Rockefeller Special Studies Report: 137
Romney, George: 112
Roosevelt, Theodore: 22-23, 25
Rousseau, Jean Jacques: 42
Royal Bank of Canada: 108
Royal Dutch Shell: 79, 138, 143
Rubber, synthetic: 79-80
Rumania, oil deposits in: 79
Rusk, Dean: 137
Russell, Bertrand: 76
Russia, oil deposits in: 79

Sacred Grove: 14
Saigon, American banks in: 82
Les Salaries Au Pouvoir!: 180
Sales, retail: 103
Salvation: 16-17
Salvatori, Henry: 136
San Francisco, taxes of: 97
Santa Barbara oil spill: 124
Saturday Review: 127
Scarcity: 162
Schultz, George: 130
Science: 117
Science: support of, 86; utilitarian, 17
Scientific American: 7
Seaborg, Glenn: 91
Security First National Bank of New York: 134
Self-determination: 157, 175; and biological liberation, 180; strategy for, 170-71; and tax structure, 174; in U.S., 166-72, 180

Selye, Hans: 24
Senate: Appropriations Committee of, 92; Armed Forces Committee of, 92
Senate Armed Forces Service Committee, Electronics Battlefield Subcommittee of: 142
Service systems, breakdown of: 10
Seskin, Eugene P.: 117
Shell Oil: 130, 143
Shenyang: 153-54
Sherrill, Robert G.: 131
Siberian Arctic, oil and gas in: 144
Sierra Club: 24, 30-31, 42, 93, 134
Sinclair Oil Corporation: 132
Singer, Fred S.: 65
Skidmore, Owens, and Merrill: 61
Small Business Committee, Senate: 130
Smith, Adam: 16, 41
Smoke particles: 69
Social balance: 160
Socialism: ecological stability and, 159; and limits on production, 161
Socialist economy, transition to: 176
"Socialist man": 151
Socialist revolution: 165
Social liberation, global: 180
Social organizations, mechanisms of: 175
Social structure, irrationality in: 4
Society: arms spending and, 95; Asian, 69; capitalist, 147-48; class nature of American, 28; control of, 50-51; disorder of, 147; division of labor and, 15; fragmentation of, 74; hierarchical structure of, 147; organization of, 27, 74; socialist, 148
Solar energy: 125
Solvents: 110
Sources: 71
Southeast Asia: American petroleum investment in, 143; petroleum reserves of, 140-41
South Vietnam: American petroleum exploration in, 141; bombing of, 90; military spending in, 143, offshore oil concessions in, 141. *See also* Indochina
Southeast Asia: American petroleum investment in, 143; petroleum reserves of, 140-41
Southwest Center for Advanced Studies: 64
Soviet Union: ecological balance in, 148-49; and Japanese petroleum market, 144; and the Middle East, 140; military spending of, 80; oil and gas discoveries in, 144
SST: 118
Standardization: 74
Standard Oil: 143; refining capacity of, 110
Standard Oil of California: 94, 124, 138-39
Standard Oil of New Jersey: 58-59; 79-80, 111, 121, 124, 130, 132-33, 137-39, 144-45; annual budget of, 135; and Cuba, 137; and Guatemala, 137; and Iran, 137; marketing brand names of, 135
Stanford University: 149
States, tax revenues of: 114
Steel industry: 102
Stern, Philip M.: 45
Sternglass, Ernst J.: 84
Stevenson, Robert: 99
Stewart, Dr. William H.: 163
Strategy for Labor: 74
Sullivan and Cromwell: 137
Sulphur oxide: 69
Sweezy, Paul: 73
Synthetic fibers: 110
Syracuse University: 124
Szilard, Leo: 86

A Tale of Two Cities: 11
Tamplin, Arthur: 84-85
Tanzer, Michael: 135
TAPS Alaskan pipeline: 129
Tax: corporate, 39; corporate income, 131; Federal, 97; petroleum, 131; special motor user, 114
Taxation, social: 75
Tax incentives: 38

Tax revolts: 174
Technics and Civilization: 171
Technology, dependence on: 51
Tehran: 135
Telephones: 60
Temperature, as energy source: 126
Terror: 83
Texaco: 124, 138
Thailand, petroleum deposits in: 141
Thales: 13
Third World: 48; American oil interests in, 99. *See also* Developing countries
Thoreau, Henry: 22
Thompkins, Paul: 85
Trace metals: 37-38
Traffic, long-distance air: 119
Train, Russell: 35, 115
Train, high-speed Japanese: 120
Trans Alaska Pipeline proposal: 133
Transportation: 118; automobiles and, 105; collective forms of, 74; diversification of, 120; monopoly in, 119; privatization of, 158
The Triumph of Conservatism: 26
Truman, Harry S: 85, 137; administration of, 45
Tydings, Senator Joseph: 36

Underdeveloped countries. *See* Developing countries, Third World
Uniformity: 77
Union Oil Company: 143
Unions: 104
United Arab Republic (UAR), poison gas use by: 87
United Auto Workers: 104
United Nations Economic Commission for Asia and the Far East (ECAFE): 141
United Nations Yearbook: 62
United States: air superiority, 92; consumption of, 65; cultural revolution in, 179; military spending of, 80
U.S. Department of Agriculture: 44, 67
U.S. Department of Defense: 78, 81, 93, 97, 99
U.S. Department of Health, Education, and Welfare: 37
U.S. Department of the Interior: 127
U.S. Department of Transportation: 115
U.S. Department of the Treasury: 129
U.S. State Department, and oil industry: 138
U.S. Steel: 107
U.S. Supreme Court: 140
U.S. Trust Company of New York: 138
University of California: 124
University of Pittsburgh: 84, 124
University of Texas: 124
Uranium: 121
Urban crises: 173
Urban environment, Federal programs and: 40
Urbanization: 62, 146-47
Utility: 14; nature and, 24

Vanderbilt University: 124
Venezuela: 135, 138
Vermont Free Territory: 180
Vietnam: 83; postwar, 92; rural culture of, 91. *See also* Indochina, South Vietnam

War: as business impetus, 79; petroleum industry and, 78-79;
Warefare: CBW, 87; chemical, 91; ecological, 90
War machine: 78
War of pollution: 5
War policy, of U.S. administration: 32
Waste: in China, 152; production of, 45; in U.S., 34, 59
Wastes: atomic, 68; crude petroleum, 68; hazardous, 65
Water treatment plants: 33; funds for (table), 34; inspection of, 37; standards of, 37
Waterways, emissions into: 35
Watson, James: 86

Wayne State University: 116
Weapons: biological, 87; chemical, 87; development of, 88
Weber, Max: 50
Wehrmacht: 79
Welfare, social: 71
Western Pacific, petroleum reserves of: 140
West Germany: 59, 147, 178
White-collar classes, commuting of: 9-10
Whiteside, Thomas: 67, 91
Whitman, Walt: 20, 22
Wilderness Society: 30
Williams, Calvin: 115
Williams, Charles: 115
Williams College: 124
Wilson, Charles: 112, 136
Wind, as energy source: 125
Winthrop, John: 18, 22
Wolff, Robert Paul: 21
Women: status of: 19, 22, 147
Woodwell, George M.: 67
Workweek: 179
World War II, employment during: 95

Yale University: 124
Yugoslavia, workers' control in: 156

Zero Population Growth: 30